Target

Get back on track

GRADE 5

D1079484

Edexcel GCSE (9-1)
Mathematics
Number and Algebra

Katherine Pate

Pearson

Published by Pearson Education Limited, 80 Strand, London, WC2R ORL.

www.pearsonschoolsandfecolleges.co.uk

Text © Pearson Education Limited 2016
Typeset by Tech-Set Ltd, Gateshead
Original illustrations © Pearson Education Ltd 2016

The right of Katherine Pate to be identified as author of this work has been asserted by her in accordance with the Copyright, Designs and Patents Act 1988.

First published 2016

20
10 9 8 7 6

British Library Cataloguing in Publication Data
A catalogue record for this book is available from the British Library

ISBN 978 0 435 18333 2

Printed in Italy by Lego S.p.A

Helping you to formulate grade predictions, apply interventions and track progress.

Any reference to indicative grades in the Pearson Target Workbooks and Pearson Progression Services is not to be used as an accurate indicator of how a student will be awarded a grade for their GCSE exams.

You have told us that mapping the Steps from the Pearson Progression Maps to indicative grades will make it simpler for you to accumulate the evidence to formulate your own grade predictions, apply any interventions and track student progress.

We're really excited about this work and its potential for helping teachers and students. It is, however, important to understand that this mapping is for guidance only to support teachers' own predictions of progress and is not an accurate predictor of grades.

Our Pearson Progression Scale is criterion referenced. If a student can perform a task or demonstrate a skill, we say they are working at a certain Step according to the criteria. Teachers can mark assessments and issue results with reference to these criteria which do not depend on the wider cohort in any given year. For GCSE exams however, all Awarding Organisations set the grade boundaries with reference to the strength of the cohort in any given year. For more information about how this works please visit: https://qualifications.pearson.com/en/support/support-topics/results-certification/understanding-marks-and-grades.html/Teacher

Each practice question features a Step icon which denotes the level of challenge aligned to the Pearson Progression Map and Scale.

To find out more about the Progression Scale for Maths and to see how it relates to indicative GCSE 9–1 grades go to www.pearsonschools.co.uk/ProgressionServices

Contents

Useful formulae

Unit 1 Mixed numbers

Area of a triangle $= \dfrac{\text{base } (b) \times \text{perpendicular height } (h)}{2}$

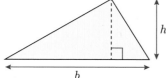

Unit 3 Formulae

Density $(D) = \dfrac{\text{mass } (m)}{\text{volume } (V)}$

Speed $(s) = \dfrac{\text{distance } (d)}{\text{time } (t)}$

$v^2 = u^2 + 2as$

where
   v is final velocity
   u is initial velocity
   a is acceleration
   s is distance

$s = ut + \frac{1}{2}at^2$

where
   s is distance
   u is initial velocity
   t is time
   a is acceleration

Unit 7 Pythagoras' theorem

$c^2 = a^2 + b^2$

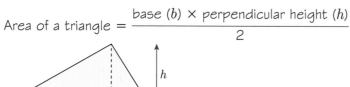

Glossary

Unit 1 Mixed numbers

Mixed numbers have a whole-number part and a fraction part.

The **lowest common multiple (LCM)** of two numbers is the smallest number that is a multiple of both numbers.

The **numerator** is the top number of a fraction.

The **denominator** is the bottom number of a fraction.

An **improper fraction** has a numerator that is bigger than its denominator.

The **reciprocal** of a fraction is the 'upside down' or **inverse** of the fraction.

Unit 2 Brackets

A **term** is a number, a letter, a number and a letter multiplied together, or two or more letters multiplied together.

The letters are called **variables** because their values can change.

The **coefficient** is the number in front of the variable.

An **algebraic expression** can contain variables, numbers and operations.

The **highest common factor (HCF)** of two numbers is the largest number that is a factor of both numbers.

Double brackets are two sets of brackets, each containing an expression, that are multiplied together.

To **expand** brackets multiply each term inside the first set of brackets by the term outside the brackets or by each term in every other set of brackets.

Factorising brackets is the reverse of expanding.

A **quadratic expression** always contains a squared term. It cannot have a power higher than 2.

The **difference of two squares** is a quadratic expression with two squared terms, and one term is subtracted from the other.

Unit 3 Formulae

A **formula** is a general rule that shows a relationship between variables.

An **equation** contains an unknown number (a letter) and an equals sign.

Derive means to work out from the information that is given.

The **subject** of a formula is the letter on its own, on one side of the equals sign.

Unit 4 Equations

To **solve** an equation, work out the value of the unknown number.

The **angles in a triangle** add up to 180°.

The **angles in a quadrilateral** add up to 360°.

A **quadrilateral** is a 2D shape with four straight sides.

An **isosceles triangle** has two equal angles and two equal length sides.

A **scalene triangle** has three different angles and 3 different length sides.

An **equilateral triangle** has three equal angles and three equal length sides.

The **perimeter** is the total distance around the edge of a shape.

Unit 5 Simultaneous equations

Simultaneous equations are equations in two (or more) variables that are true at the same time.

In algebra, **substitute** means to put numbers in place of letters.

Unit 6 Indices

In index notation, the number that is multiplied by itself is called the **base**.

In index notation, the number written above the base is called the **index** or the **power**.

The **index** tells you how many times the base must be multiplied by itself.

You can round numbers to a certain number of **significant figures (s.f.)**. The first significant figure is the one with highest **place value**. It is the first non-zero digit in the number, counting from the left.

The **laws of indices** for multiplication and division are the rules for performing these operations. To multiply powers of the same number, add the indices. To divide two powers of the same number, subtract the indices.

Unit 7 Pythagoras' theorem

Pythagoras' theorem shows the relationship between the lengths of the three sides of a right-angled triangle. When the two shorter sides are of lengths a and b, the length of the longest side (the **hypotenuse**) is given by $c^2 = a^2 + b^2$.

Coordinates are values on the x- and y-axes on a graph.

Unit 8 The equation of a straight line

The **y-axis** is the vertical axis on a graph.

The **x-axis** is the horizontal axis on a graph.

The steepness of a graph is called the **gradient**.

The **y-intercept** is where a graph crosses the y-axis.

The **equation of a straight line** with gradient m and y-intercept c is $y = mx + c$.

Parallel lines are always the same distance apart and never meet.

Unit 9 Non-linear graphs

A **cubic equation** contains a term in x^3 and no higher power of x.

Turning points are where a graph changes direction and gradient:

maximum turning point ∩

minimum turning point ∪

A **line of symmetry** on a graph is a line that divides the graph into two halves, where one half is a reflection of the other.

The **x-intercept** is where a graph crosses the x-axis.

Unit 10 Inequalities

An **inequality** is a relationship between two values that are different, such as that an expression is greater than a number, or that a variable is less than or equal to a number.

Inequalities have a **solution set** that is a range of values.

An **integer** is a positive or negative whole number, or zero.

Unit 11 Direct and inverse proportion

When two or more quantities are in **direct proportion**, they increase or decrease by the same scale factor.

When two quantities are in **inverse proportion**, as one increases, the other decreases in the same ratio.

$y \propto x$ means y is proportional to x. When $y \propto x$, then $y = kx$, where k is the constant of proportionality.

$y \propto \dfrac{1}{x}$ means x and y are in inverse proportion.

① Mixed numbers

This unit will help you add, subtract, multiply and divide mixed numbers.

AO1 Fluency check

① Work out

a $\frac{1}{4} + \frac{3}{8} =$ $\frac{20}{32}$ **b** $\frac{2}{3} - \frac{1}{6} =$ $\frac{9}{18}$ $\frac{1}{2}$ **c** $\frac{2}{7} + \frac{10}{3} =$ $\frac{30}{21}$ **d** $\frac{15}{8} - \frac{2}{5} =$ $\frac{59}{40}$

② Work out

a $\frac{2}{3} \times \frac{1}{5} =$ $\frac{2}{15}$ **b** $\frac{5}{7} \div \frac{2}{3} =$ **c** $\frac{2}{5} \times \frac{4}{3} =$ $\frac{8}{15}$ **d** $\frac{7}{8} \div \frac{3}{2} =$

③ Complete

a $3\frac{1}{4} = \frac{13}{4}$ **b** $2\frac{5}{6} = \frac{17}{6}$ **c** $4\frac{2}{5} = \frac{22}{5}$ **d** $5\frac{1}{3} = \frac{16}{3}$

e $\frac{8}{3} = 2\frac{2}{3}$ **f** $\frac{11}{7} = 1\frac{4}{7}$ **g** $\frac{22}{5} = 4\frac{3}{5}$ **h** $\frac{15}{4} = 3\frac{3}{4}$

④ **Number sense**

$\boxed{\frac{1}{3} \times 42 = 14}$ Use this number fact to work out

a $\frac{2}{3} \times 42 =$ **b** $\frac{4}{3} \times 42 =$ **c** $\frac{1}{3} \times 84 =$

d $\frac{1}{3} \times 21 =$ **e** $14 \div 42 =$ **f** $14 \div \frac{1}{3} =$

Key points

Mixed numbers have a whole-number part and a fraction part.

To add mixed numbers, write them as improper fractions first.

These **skills boosts** will help you calculate with mixed numbers.

> **1** Adding and subtracting mixed numbers
> **2** Multiplying mixed numbers
> **3** Dividing mixed numbers

You might have already done some work on mixed numbers. Before starting the first skills boost, rate your confidence working out these calculations.

1 $2\frac{3}{4} + 1\frac{1}{2}$

$\frac{4}{4} + \frac{3}{2}$

$\frac{34}{8}$ $4\frac{2}{8}$

2 $3\frac{1}{2} \times \frac{1}{4}$

$\frac{7}{2} \times \frac{1}{4}$

$\frac{7}{8}$

3 $2\frac{2}{3} \div \frac{1}{2}$

How confident are you?

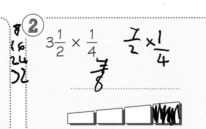

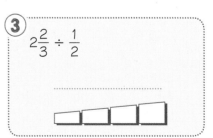

1 Adding and subtracting mixed numbers

Guided practice

Work out

$2\frac{1}{4} + 3\frac{1}{2}$ $\frac{9}{4} + \frac{7}{2}$ $\frac{46}{8}$

Write as improper fractions.

$= \frac{9}{4} + \frac{\cdots}{2}$

Use a bar model to help you convert the mixed numbers to improper fractions.

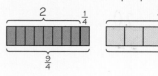

Find the lowest common multiple (LCM) of 4 and 2.

$= \frac{9}{4} + \frac{\cdots}{4}$

The LCM of 4 and 2 is 4.

Add the numerators of the fractions.

$= \frac{\cdots}{4}$

Write as a mixed number.

$= 5\frac{3}{4}$

1 Work out

a $2\frac{3}{4} + 1\frac{5}{8} = $ ....

b $4\frac{2}{5} + 1\frac{5}{6} = $ ....

c $1\frac{2}{5} + 3\frac{1}{7} = $ ....

2 Work out

a $3\frac{5}{8} - 1\frac{1}{4} = $ ....

b $4\frac{2}{3} - 2\frac{1}{6} = $ ....

c $5\frac{3}{8} - 3\frac{1}{3} = $ ....

Exam-style question

3 Work out **a** $1\frac{3}{4} + \frac{2}{5} = $

.... (2 marks)

b $3\frac{4}{7} - 1\frac{1}{2} = $

.... (2 marks)

4 Work out

a $6 - 2\frac{5}{8} = $ ....

b $5 - 3\frac{2}{5} = $ ....

c $4 - 2\frac{5}{9} = $ ....

5 Here are the weights of three puppies, in kg.

$4\frac{3}{4}, \quad 3\frac{5}{8}, \quad 4\frac{1}{2}$

Calculate their total weight.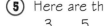

Reflect In Q4 you subtracted mixed numbers from whole numbers by writing them as improper fractions. Is this the best method to use? If not, write a calculation to show the method you like best.

2 Multiplying mixed numbers

Guided practice

Work out

$1\frac{1}{2} \times \frac{3}{4}$

Write as improper fractions.

$= \frac{}{2} \times \frac{}{4}$

Multiply numerators and denominators.

$= \frac{3 \times 3}{2 \times 4}$

$= \frac{}{8}$

Write as a mixed number.

$= 1\frac{1}{8}$

The numerator is the 'top' of the fraction. The denominator is the 'bottom'.

$1\frac{1}{2} \times \frac{3}{4}$ means $1\frac{1}{2}$ lots of $\frac{3}{4}$

1 of $\frac{3}{4}$ = $\frac{6}{8}$

$\frac{1}{2}$ of $\frac{3}{4}$ = $\frac{3}{8}$ } $\frac{9}{8}$

(1) Work out

a $1\frac{3}{4} \times \frac{5}{6} =$ ....

b $2\frac{1}{2} \times \frac{3}{8} =$ ....

c $\frac{2}{3} \times 1\frac{4}{5} =$ ....

(2) Work out

a $1\frac{1}{2} \times 1\frac{3}{4} =$ ....

b $2\frac{1}{5} \times 1\frac{1}{4} =$ ....

c $3\frac{2}{3} \times 5\frac{1}{2} =$ ....

Exam-style question

(3) Work out the area of this rectangle in square inches.

$1\frac{5}{8}$ inches

$2\frac{3}{4}$ inches

.... (2 marks)

(4) Calculate the area of this triangle.

$1\frac{1}{2}$ m

$3\frac{3}{4}$ m

Give your answer in square metres. ....

Reflect Maria says, 'It's best to start all fraction calculations by changing to improper fractions first.' Do you agree with Maria?

3 Dividing mixed numbers

Dividing by a number is the same as multiplying by its reciprocal.

Guided practice

Worked exam question

Work out

$$3\frac{1}{4} \div 2\frac{1}{2}$$

Write as improper fractions.

$$= \frac{13}{4} \div \frac{\text{.........}}{2}$$

Multiply by the reciprocal.

$$= \frac{13}{4} \times \frac{2}{\text{.........}}$$

The reciprocal of $\frac{5}{2}$ is $\frac{2}{5}$

$$= \frac{26}{\text{.........}}$$

Write the improper fraction as a mixed number.

$$= 1\frac{6}{\text{.........}}$$

Simplify

$$= 1\frac{3}{\text{.........}}$$

(1) Work out

a $1\frac{1}{3} \div \frac{5}{8} =$ ....

b $2\frac{1}{5} \div \frac{3}{5} =$ ....

c $3\frac{2}{5} \div 1\frac{1}{4} =$ ....

(2) Work out

a $3\frac{1}{4} \div 1\frac{1}{2} =$ ....

b $5\frac{1}{6} \div 2\frac{1}{3} =$ ....

c $1\frac{3}{4} \div 2\frac{1}{2} =$ ....

Exam-style question

(3) Find the length of the side labelled x, in centimetres.

Area $= 4\frac{3}{8}$ cm²

x

$3\frac{1}{2}$ cm

.... (2 marks)

Reflect

What fraction skills have you used in this skills boost?

Practise the methods

Answer this question to check where to start.

Check up

Tick the calculation equivalent to $3\frac{1}{4} + 1\frac{3}{5}$.

A ◯

$$\frac{31}{4} + \frac{13}{5} = \frac{207}{20}$$

B ◯

$$\frac{13}{4} + \frac{8}{5} = \frac{21}{9}$$

C ◯

$$\frac{13}{4} + \frac{8}{5} = \frac{97}{20}$$

If you ticked C go to Q4.

If you ticked A or B go to Q1 for more practice.

① $1\frac{3}{5}$ **a** How many $\frac{1}{5}$s in 1? ....   **b** How many $\frac{1}{5}$s in $1\frac{3}{5}$? ....

② $3\frac{1}{4}$ **a** How many $\frac{1}{4}$s in 3? ....   **b** How many $\frac{1}{4}$s in $3\frac{1}{4}$? ....

③   **a** $\frac{5}{4} + \frac{9}{8} =$ ....   **b** $\frac{11}{6} + \frac{5}{3} =$ ....   **c** $\frac{9}{5} + \frac{7}{2} =$ ....

④   **a** $1\frac{1}{3} \div \frac{5}{8} =$ ....   **b** $2\frac{1}{5} \div \frac{3}{5} =$ ....   **c** $3\frac{2}{5} \div 1\frac{1}{4} =$ ....

⑤   **a** $3\frac{5}{6} + \frac{2}{3} =$ ....   **b** $1\frac{1}{4} + 2\frac{7}{8} =$ ....   **c** $2\frac{1}{5} + 1\frac{3}{10} =$ ....

⑥   **a** $1\frac{2}{5} + 3\frac{1}{4} =$ ....   **b** $2\frac{5}{7} + 1\frac{1}{2} =$ ....   **c** $4\frac{5}{8} + \frac{5}{12} =$ ....

⑦   **a** $4\frac{3}{5} - 2\frac{1}{10} =$ ....   **b** $5\frac{5}{6} - 3\frac{2}{3} =$ ....   **c** $3\frac{7}{8} - 2\frac{1}{2} =$ ....

⑧   **a** $5\frac{3}{5} - 2\frac{1}{8} =$ ....   **b** $3\frac{4}{7} - 2\frac{1}{6} =$ ....   **c** $4\frac{3}{8} - \frac{5}{6} =$ ....

⑨   **a** $3\frac{1}{2} \times 1\frac{1}{2} =$ ....   **b** $2\frac{1}{2} \times 1\frac{1}{3} =$ ....   **c** $1\frac{3}{5} \times 2\frac{3}{4} =$ ....

⑩   **a** $2\frac{2}{3} \div \frac{4}{5} =$ ....   **b** $1\frac{2}{3} \div 2\frac{1}{2} =$ ....   **c** $1\frac{3}{5} \div 1\frac{1}{7} =$ ....

Exam-style question

⑪ Work out

   a $1\frac{5}{8} - \frac{1}{6} =$ .... **(2 marks)**

   b $3\frac{2}{5} \div 1\frac{1}{2} =$ .... **(2 marks)**

Problem-solve!

1 A maths textbook is $1\frac{7}{8}$ inches thick.

The shelf in the maths classroom is $34\frac{1}{2}$ inches long.

Will a set of 20 books fit on the shelf?
Show working to explain.

$1\frac{7}{8}$ inches

.... (3 marks)

2 Amina cuts $4\frac{5}{8}$ yards of ribbon into three equal-length pieces.

Work out the length of one piece.
Give your answer as a mixed number of yards.

....

3 Rebecca screws a washer and a nut on to a bolt, as shown in the diagram.
What length of the bolt sticks out from the nut?

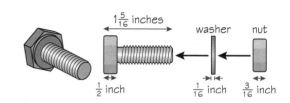

$1\frac{5}{16}$ inches
washer   nut
$\frac{1}{2}$ inch $\frac{1}{16}$ inch $\frac{3}{16}$ inch

....

4 Pat has $6\frac{1}{2}$ bags of sand, which weigh $3\frac{1}{4}$ kg each.

a Calculate their total weight.   ....

b Pat's bucket holds $2\frac{1}{2}$ kg of sand.

How many full bucket loads of sand does he have? ....

5 A photograph $5\frac{1}{2}$ inches by 7 inches is in a frame $\frac{5}{8}$ inch wide.

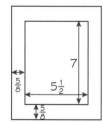

Calculate the total area of the photograph and the frame.
Give your answer as a fraction.

.... (4 marks)

6 You can use these digits.

| 1 | 2 | 3 | 4 | 5 | 6 |

Write the digits in this mixed number calculation to give the smallest possible answer.
Use each digit only once.

$$\ldots\ldots\frac{\ldots\ldots}{\ldots\ldots} \div \ldots\ldots\frac{\ldots\ldots}{\ldots\ldots} = \ldots\ldots$$

Now that you have completed this unit, how confident do you feel?

1 Adding and subtracting mixed numbers

2 Multiplying mixed numbers

3 Dividing mixed numbers

② Brackets

This unit will help you factorise expressions and expand double brackets.

AO1 Fluency check

① Expand

  a $2(x + 3) =$ $2x + 6$ **b** $-4(x + 1) =$ $-4x-4$ **c** $x(x + 3) =$ $x^2 + 3x$

② Simplify

  a $2x + 4x =$ $6x$ **b** $x^2 + 4x - 2x =$ $3x^2$ **c** $x^2 - 5x + 2x - 3 =$ _____

③ Find the highest common factor (HCF) of

  a x and $2x$ _____   **b** xy and $4x$ _____   **c** x^2 and x _____

④ Find the missing terms.

  a $x \times$ _____ $= 3x$ **b** $x \times$ _____ $= xy$ **c** $x \times$ _____ $= x^2$

⑤ Factorise

  a $4x + 12 =$ _____   **b** $2x + 10 =$ _____   **c** $4x - 6 =$ _____

⑥ Number sense

  Find the missing values.

  a $3 \times$ _____ $= 15$ **b** $3 \times$ _____ $= -15$ **c** $-3 \times$ _____ $= 15$

Key points

To factorise an expression find the HCF of the terms.	The HCF may be a letter term. For example, the HCF of $3x$ and xy is x.	3^2 means 3×3 x^2 means $x \times x$ $(x + 3)^2$ means $(x + 3)(x + 3)$

These **skills boosts** will help you factorise different types of expressions and check your answer by expanding.

1 Factorising when the HCF is a letter or a number 〉 **2** Factorising when the HCF has numbers and letters 〉 **3** Expanding double brackets 〉 **4** Factorising quadratic expressions

You might have already done some work on factorising and expanding. Before starting the first skills boost, rate your confidence factorising or expanding each type of expression.

1
Factorise
$3x + xt$

2
Factorise
$4x^2 + 8x$

3
Expand
$(x + 2)(x + 5)$

4
Factorise
$x^2 + 4x + 3$

How confident are you?

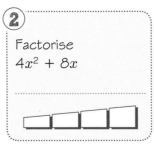
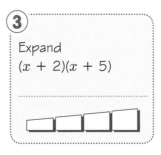
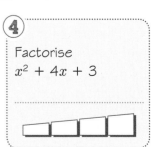

1 Factorising when the HCF is a letter or a number

Guided practice

 Worked exam question

Factorise $2x + xy$

Find the HCF of $2x + xy$.

The HCF of $2x$ and xy is ....   The HCF may be a letter term.

Write the HCF outside the bracket.

$2x + xy = x($ + $)$

Find the terms inside the bracket.

$2x + xy = x(\text{......} + \text{......})$

Check your answer by expanding.

$\qquad = x(2 + y)$

$\qquad = \text{............} + xy$

[diagram]
$$\begin{array}{c|c|c} & 2 & y \\ \hline x & x \times 2 = 2x & x \times y = xy \end{array}$$
$2 + y$

Expand $\longrightarrow$

$x(2 + y) = 2x + xy$

$\longleftarrow$ Factorise

(1) Factorise   **Hint** $y \times 1 = y$

   a $6x + xt =$ ....   **b** $5y - xy =$ ....   **c** $y^2 + y =$ ....

   d $4x + xy =$ ....   **e** $5x + x^2 =$ ....   **f** $3a - ab =$ ....

(2) Factorise

   a $x^2y + x$ **b** $c^2d + cd$ **c** $st^2 - st$

     $= x(\text{......} + \text{......})$ $= cd(\text{......} + \text{......})$ $=$ ....

(3) Factorise

   a $x^2y + xy^2$ **b** $yz^2 - y^2z$ **c** $a^2b^3 + a^2b$

     $= xy(\text{......} + \text{......})$ $=$ .... $= a^2b(\text{......} + \text{......})$

   d $d^2e + d^2f$ **e** $u^3v^2 - u^3v$ **f** $g^2h^2 + g^3h^2$

     $=$ .... $=$ .... $=$ ....

Exam-style question

(4) a Factorise $\quad 5x + 20$ .... (1 mark)

   b Factorise $\quad x^2 - 3x$ .... (1 mark)

   c Factorise $\quad x^2y + xy$ .... (2 marks)

Reflect What is the best way to check whether your factorisation is correct?

2 Factorising when the HCF has numbers and letters

The HCF may be a number and letter term. The HCF of $2x$ and $4xy$ is $2x$.
'Factorise completely' means 'write the HCF outside the bracket'.

Guided practice

Factorise

a $3x^2 + 6x$

b $5x^2y - 10xy$

---

a Find the HCF of $3x^2 + 6x$.

Write the HCF outside the bracket.

$3x^2 + 6x =$ .... (+)

Find the terms inside the bracket.

$3x^2 + 6x = 3x ($ .... + ... $)$

Check your answer by expanding

$= 3x(x + 2)$

$=$ .... + ...

Find the HCF of the numbers and the HCF of the letters.

$$
\begin{array}{c|c|c}
 & x & 2 \\
\hline
3x & 3x \times 2 = 3x^2 & 3x \times 2 = 6x \\
\end{array}
$$

with $x + 2$ across the top.

b Find the HCF of $5x^2y$ and $10xy$.

$5x^2y - 10xy =$ .... (-)

Find the terms inside the bracket.

$5x^2y - 10xy = 5xy ($ .... - .... $)$

Check your answer by expanding

$= 5xy(x - 2)$

$=$ .... - ....

$5xy \times$ .... $= 5x^2y$

$5xy \times$ .... $= 10xy$

---

(1) Factorise completely

a $4x^2 + 8x$

b $3bc - 12b^2$

c $6x^2 + 4xy$

d $6x^2 - 3x$

e $5m^2 - 20m$

f $2ad + 10d^2$

② **a** $7pq - 14p^2 + 14pq^2$

b $3y + 15x^2y^2 - 9xyz$

③ Explain why these are not complete factorisations of $8x^2 - 12xy$.

Hint $4x^2$ and $-6xy$ have two common factors, 2 and ....

a $2(4x^2 - 6xy)$

b $2x(2x - 6y)$

c $4(x^2 - 3xy)$

④ Factorise completely

a $8x^2 - 12xy$

b $5abc + 15a^2c$

c $9xy + 27xy^2$

d $6e^2f - 18efg$

e $20x^2y - 30xy^2$

f $6ab^2 - 30b^2$

⑤ **a** $8w^2z^2 - 24v^2wz + 16vw^2z$

b $6ab^2c - 3ab^2c^2 - 12b^2c$

Exam-style question

⑥ Factorise completely

a $8x - 16$ .... (1 mark)

b $6xy^2 + 12x^2y^2$ .... (2 marks)

Reflect How can you remember what 'expand' and 'factorise' mean?

10   Unit 2 Brackets

3 Expanding double brackets

Worked exam question

Guided practice

Expand $(x + 3)(x + 5)$

Split $(x + 3)(x + 5)$

into $x(x + 5) + 3(x + 5)$

Expand the brackets.

$$= x^2 + 5x + \text{\_\_\_\_\_} + \text{\_\_\_\_\_}$$

Collect like terms.

$$= x^2 + \text{\_\_\_\_\_} + 15$$

Work out x lots of $x + 5$ and 3 lots of $x + 5$.

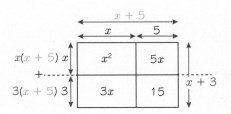

① Expand and simplify

a $(x + 2)(x + 4)$ **b** $(x + 3)(x - 1)$ **c** $(x - 2)(x + 5)$

d $(x - 3)(x + 2)$ **e** $(m - 2)(m - 7)$ **f** $(y - 4)(y - 1)$

Exam-style question

② Expand and simplify $(m + 5)(m + 8)$.   .... (2 marks)

③ Expand and simplify

a $(x + 1)(x - 1)$ **b** $(x + 2)(x - 2)$ **c** $(b + 5)(b - 5)$

④ Predict the expansions of

a $(x + 3)(x - 3)$ **b** $(a + 10)(a - 10)$ **Hint** Look at your answers to Q3. Check by expanding.

⑤ Expand and simplify

a $(x + 3)^2$ **b** $(x + 4)^2$ **c** $(x - 5)^2$
   $= (x + 3)(x + 3)$

⑥ Show that the area of this square is $x^2 + 4x + 4$. $x + 2$

Hint 'Show that' means
'Write some maths to show that'.

Exam-style question

⑦ Show that $x(x + 1) - x - 1 = (x + 1)(x - 1)$

(3 marks)

Reflect   Why do you think the answers in Q3 and Q4 are called 'difference of two squares'?

4 Factorising quadratic expressions

A quadratic expression like this may factorise into two brackets.

x^2 term x term   number term
$$x^2 \quad + \quad 7x \quad + \quad 10$$

Expand →

$(x + 2)(x + 5) = x^2 + 7 + 10$

← Factorise 2×5 2×5

Guided practice

Worked exam question

Factorise $x^2 + 7x + 12$.

Write the factor pairs of 12.

1×12
2×6
3×4

Find two number that multiply to give 12.

Find the factor pair that add to give 7.

$1 + 12 = 13$ ✗
$2 + 6 = 8$ ✗
$3 + 4 = 7$ ✓

Split $7x$ into $3x$ and $4x$.

$x^2 + 7x + 12 = \underbrace{x^2 + 3x} + \underbrace{4x + 12}$

$= x(x + 3) + 4(x + 3)$

$= (x + 3)(x + 4)$

Factorise pairs of terms.

① Factorise

a $x^2 + 4x + 3$ **b** $x^2 + 6x - 7$ **c** $x^2 - 5x - 6$

② Factorise   **Hint** Look for factors of -7 that add to make 6.

a $x^2 + 4x - 5$ **b** $x^2 + x - 6$ **c** $x^2 - 4x + 3$

③ Factorise

a $x^2 - 25$ **b** $x^2 - 81$

Hint Factorising difference of two squares.
$x^2 - 16 = (x - 4)(x + 4)$
$x^2 - 4 = (x - 2)(x + 2)$
Look back at Skills boost 3.

Exam-style question

④ Factorise

a $x^2 - 100$ **(2 marks)**

b $x^2 - 8x + 16$ **(2 marks)**

Reflect How can you tell that a factorisation will have negative numbers in it?

Practise the methods

Answer this question to check where to start.

Check up

Tick the correct factorisation of $4x^2 - 12xy$.

A $4(x^2 - xy)$ ○

B $2x(2x - 6y)$ ○

C $4x(x - 3y)$ ○

If you ticked C go to Q3.

If you ticked A or B go to Q1 for more practice.

1 Fill in the missing numbers in these factorisations

a $2x^2 + 4xy = 2x(x + \text{........} y)$

b $6ab + 9a^2 = 3a(\text{........} b + \text{........} a)$

2 Explain why these are not complete factorisations of $8x^2 - 12xy$.

a $2(4x^2 - 6xy)$

b $2x(4x - 6y)$

c $4(2x^2 - 3xy)$

Hint $4x^2$ and $-6xy$ have two common factors, 2 and ....

3 Factorise

a $3x + xz$

b $a^2 - a$

c $2b + b^2$

d $m^2p + mp$

e $2xy + 6x^2$

f $3cd^2 - 12c^2$

4 Factorise completely

a $10x^2 + 20xy$

b $16bc - 4c^2d$

c $12xy^3 - 6x^2y$

5 Expand and simplify

a $(x + 5)(x + 7)$

b $(n + 2)(n - 7)$

c $(t - 8)(t - 3)$

6 Expand and simplify

a $(x + 2)^2$

b $(x - 9)(x + 9)$

c $(x - 8)^2$

7 Factorise

a $x^2 + 7x + 10$

b $x^2 - x - 6$

c $x^2 - 8x + 7$

Exam-style question

8 Factorise

a $x^2 - 16$ .... (2 marks)

b $x^2 - 10x + 25$ .... (2 marks)

Problem-solve!

(1) Show that the area of this rectangle is $x^2 + 4x - 5$.

$x + 5$

$x - 1$

....

Exam-style questions

(2) Show that $x(2 + x) + 1 = (x + 1)^2$

.... (2 marks)

(3) ABCD is a rectangle.
Write an expression for the area of triangle BCD.

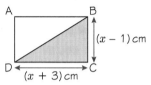

$(x - 1)$ cm

$(x + 3)$ cm

.... (3 marks)

(4) Show that $x^2 + 6x + 5 = (x + 3)^2 - 4$

.... (2 marks)

(5) Show that the shaded area is $6(2x + 7)$.

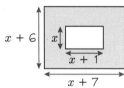

$x + 6$ x

$x + 1$

$x + 7$

.... (3 marks)

(6) In rectangle PQRS, PQ is 5 cm longer than PS.
Write an expression for the area of PQRS.

x

.... (2 marks)

(7) Write an expression for the missing length in this triangle.

$x + 2$

x

.... (3 marks)

Now that you have completed this unit, how confident do you feel?

1 Factorising when the HCF is a letter or a number

2 Factorising when the HCF has numbers and letters

3 Expanding double brackets

4 Factorising quadratic expressions

③ Formulae

This unit will help you to use, write and rearrange formulae.

① When $a = 2$, $b = 5$ and $c = -1$, find the value of

a $a + b$ ....   **b** ab ....   **c** $\dfrac{c}{a}$ ....   **d** a^2 ....   **e** $\sqrt{8a}$ ....   **f** b^2c ....

② Work out

a $3^2 - 5 =$ ....   **b** $\dfrac{8 + 2}{5} + \sqrt{9} =$ ....

③ Solve to find x.

a $4x - 1 = 19$ ....   **b** $\dfrac{x}{3} = 2$ ....

④ A box contains x counters. Write an expression for the number of counters in

a 3 boxes   **b** n boxes   **c** $\dfrac{1}{2}$ a box   **d** 2 boxes plus 10 extra counters.

⑤ **Number sense**

Complete these calculations.

$7 \times$ .... $= 56$ $9 \times$ .... $= 45$

$56 \div$ .... $= 8$ $45 \div$ .... $= 5$

$56 \div$ .... $= 7$ $45 \div$ .... $= 9$

Key points

| A formula connects two or more quantities. | You can use letters for the variables. | To use a formula, substitute (swap) the values you are given for the letters. |

These **skills boosts** will help you to use, create and rearrange formulae.

1 Finding a value that is not the subject of a formula ▸ **2** Deriving a simple formula ▸ **3** Changing the subject of a formula ▸ **4** Changing the subject of a more complex formula

You might have already done some work on formulae. Before starting the first skills boost, rate your confidence in each skill.

①

$v = u + at$

Find t when $v = 10$, $u = 0$ and $a = 5$

②

Maria's pay is £8 per hour. Write a formula connecting pay and hours worked.

③

Make d the subject of $s = \dfrac{d}{t}$

④

Make A the subject of $P = \dfrac{F}{A}$

How confident are you?

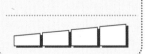

1 Finding a value that is not the subject of a formula

Substitute all the values you are given into the formula. This will give you an equation to solve.

Guided practice

Worked exam question

$v = u + at$
Find u when $v = 20$, $t = 3$ and $a = 4$

_____

$$v = u + at$$

Write the values instead of the letters.

$at = a \times t$

.... $= u +$ .... $\times$ ....

Simplify

.... $= u +$ ....

Solve to find u

.... $-$ .... $= u$

$$u = 8$$

① $V = IR$

a Find R when $V = 18$ and $I = 6$

b Find I when $V = 16$ and $R = 8$

....

....

② $D = \dfrac{m}{V}$

a Find m when
$D = 8$ and $V = 5$

b Find V when
$D = 13$ and $m = 39$

c Find V when
$D = 12$ and $m = 6$

....

....

....

③ $A = \dfrac{1}{2}bh$

a Find b when
$A = 12$ and $h = 6$

b Find b when
$A = 2$ and $h = 0.5$

c Find h when
$A = 12.5$ and $b = 10$

....

....

....

Exam-style question

Hint speed $= \dfrac{\text{distance}}{\text{time}}$

④ A train travels at an average speed of 60 miles per hour.

How long does the train take to travel 210 miles? ....

(3 marks)

⑤ $v = u + at$

a Find t when $v = 20$, $u = 0$ and $a = 10$

b Find a when $v = 15$, $u = 7$ and $t = 4$

....

....

⑥ $v^2 = u^2 + 2as$

a Find a when $v = 8$, $u = 0$ and $s = 16$

b Find u when $v = 9$, $a = 4$ and $s = 7$

....

....

Reflect How have you used inverse operations to solve the equations?

② Deriving a simple formula

To write a formula, write some calculations and look for a pattern.
Choose letters for the quantities.

Guided practice

Child cinema tickets cost £5.60 each.
Write a formula connecting total cost and number of child tickets.

1 ticket: Total cost = 5.6 × 1

2 tickets: Total cost = ... × 2

n tickets: Total cost = ... × ....

$$T = 5.6n$$

n tickets

| £5.60 | £5.60 | ... | £5.60 |

5.6 × n can be written as 5.6n

① Adult cinema tickets cost £9.80 each.
Write a formula connecting total cost and number of adult tickets.

Hint Let m be the number of adult tickets.

② Child cinema tickets cost £5.60 each and
adult cinema tickets cost £9.80 each.
Write a formula for the total cost, T,
of adult and child tickets.

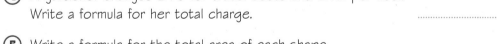

m tickets · n tickets

| £9.80 | £9.80 | ... | £9.80 | £5.60 | ... | £5.60 |

Exam-style question

③ Write a formula to connect the cost
of bike hire and number of hours.

Hint Choose letters for
cost and number of hours.

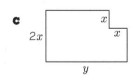

BIKE HIRE
£15
plus **£6.50** PER HOUR

.... **(2 marks)**

④ A gardener charges £10 for travel costs and £12 per hour.
Write a formula for her total charge.

⑤ Write a formula for the total area of each shape.

a x 3   4 x

b   1 x $x + 2$ x

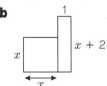

c x 2x x y

⑥ Write a formula for the total volume of this solid shape.

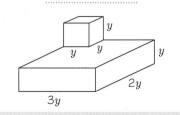

y y y y 2y 3y

Reflect   Does it matter which letters you use in a formula?

3 Changing the subject of a formula

The subject of a formula is on its own, on one side of the equals sign.

The subject of the formula $A = \frac{1}{2}bh$ is A.

Guided practice

Make h the subject of the formula $A = \dfrac{bh}{2}$

Rearrange the formula so that h is on its own, on one side of the equals sign.

$$A = \frac{bh}{2}$$

Multiply both sides by 2.

$$2 \times \ldots\ldots\ldots = \frac{{}^1\cancel{2} \times bh}{\cancel{2}}$$

$$2A = \ldots\ldots\ldots$$

Divide both sides by b

$$\frac{\ldots\ldots\ldots}{b} = \frac{{}^1\cancel{b}h}{\cancel{b}}$$

$$\frac{2A}{b} = h$$

Draw function machines with the new subject as the input.

$$h \rightarrow \boxed{\times b} \rightarrow \boxed{\div \tfrac{1}{2}} \rightarrow A$$

Work backwards with inverse operations.

$$\frac{2A}{b} \leftarrow \boxed{\div b} \leftarrow \boxed{\times 2} \leftarrow A$$

(1) **a** Make I the subject of $V = IR$

....

b Make m the subject of $D = \dfrac{m}{V}$

....

(2) **a** Make d the subject of $s = \dfrac{d}{t}$

....

b Make F the subject of $P = \dfrac{F}{A}$

....

(3) **a** Make b the subject of $A = \dfrac{1}{2}bh$

....

b Make u the subject of $v = u + at$

....

(4) **a** Make a the subject of $v = u + at$

....

b Make t the subject of $v = u + at$

....

Exam-style question

(5) $y = mx + c$

Make x the subject of this formula.   .... (2 marks)

Reflect

$8 = 2x$

$$h \rightarrow \boxed{\times 2} \rightarrow 8$$
$$4 \leftarrow \boxed{\div 2} \leftarrow 8$$

or divide both sides of the equation by ....

4 Changing the subject of a more complex formula

Identify the operations used on the variable: +, −, × or ÷.
Use the inverse operations to change the subject.

Guided practice

Make t the subject of the formula $d = \dfrac{s}{t}$

Take the reciprocal of both sides.

$$d = \frac{t}{s}$$
$$\frac{1}{d} = \frac{t}{s}$$

Multiply both sides by s.

$$\frac{s \times \cdots}{d} = \frac{{}^1\!s \times t}{\cancel{s}}$$

$$\frac{s}{d} = t$$

To get t on the top of the fraction.

Draw function machines with the new subject as the input.

$$t \rightarrow \boxed{\div s} \rightarrow \frac{1}{d}$$

Work backwards with inverse operations.

$$\frac{s}{d} \leftarrow \boxed{\times s} \leftarrow \frac{1}{d}$$

$$\frac{s}{s} = 1$$

(1) **a** Make V the subject of $D = \dfrac{m}{V}$

b Make A the subject of $P = \dfrac{F}{A}$

....

....

(2) **a** Make I the subject of $V = IR$

b Make d the subject of $s = \dfrac{d}{t}$

....

....

(3) **a** Make h the subject of the formula $A = \dfrac{bh}{2}$

b Make b the subject of $A = \frac{1}{2}bh$

....

....

(4) **a** Make F the subject of $P = \dfrac{F}{A}$

b Make A the subject of $P = \dfrac{F}{A}$

Hint Take the reciprocal first $\dfrac{1}{P} = \dfrac{A}{F}$

....

....

(5) Make t the subject of $q = \dfrac{p}{t} + x$

....

Hint Get the term with t 'on its own' on one side of the equals sign.

$$q - x = \frac{p}{t}$$

Then take the reciprocals.

$$\frac{1}{q - x} = \frac{\cdots}{\cdots}$$

(6) Make x the subject of

a $m = \dfrac{b}{x} + c$

b $y = \dfrac{r}{x} - f$

....

....

(7) Make x the subject of

a $x^2 = 25 + a$

 Hint Take the square root of both sides.

b $x^2 = d - 16$

....

c $x^2 = v + u$

d $x^2 = \dfrac{m}{n}$

....

....

(8) Make t the subject of

a $t^2 = 3m + r$

b $t^2 + k = x$

Hint Get t^2 'on its own' first.

....

....

c $t^2 - \dfrac{x}{3} = y$

d $\dfrac{t^2}{s} = n$

....

....

Exam-style question

(9) Make u the subject of the formula $v^2 = u^2 + 2as$

.... **(2 marks)**

(10) Make y the subject of

a $\sqrt{y} = x$

Hint The inverse of $\sqrt{\ }$ is 'square'.

b $\dfrac{\sqrt{y}}{4} = t$

....

....

c $n\sqrt{y} = p$

d $\sqrt{\dfrac{y}{5}} = q$

Hint $\left(\sqrt{\dfrac{y}{5}}\right)^2 = \dfrac{y}{5}$

....

....

Reflect What is the inverse operation of squaring?

Practise the methods

Answer this question to check where to start.

Check up

Make r the subject of $p = \dfrac{r}{s} + t$

Tick the correct answer.

A ○
$r = p - t \times s$

B ○
$r = s(p - t)$

C ○
$r = p - st$

| If you ticked B go to Q2. | If you ticked A or C go to Q1 for more practice. |

(1) Complete the function machines. Start at the right-hand end each time.

a $2(\ldots\ldots\ldots) \leftarrow \boxed{\times 2} \leftarrow x + 3 \leftarrow \boxed{+ 3} \leftarrow x$

b $3(\ldots\ldots\ldots) \leftarrow \boxed{\times 3} \leftarrow \ldots\ldots\ldots \leftarrow \boxed{+ k} \leftarrow m$

c $a(\ldots\ldots\ldots) \leftarrow \boxed{\times a} \leftarrow \ldots\ldots\ldots \leftarrow \boxed{+ y} \leftarrow t$

(2) Make z the subject of $f = \dfrac{z}{t} - m$ ....

(3) Make x the subject of $r = qx^2$ ....

(4) $d = st$

    a Find s when $d = 18$ and $t = 6$ **b** Find t when $d = 22$ and $s = 4$

....

Exam-style question

(5) $P = 2x + 5y$

Find x when $P = 6$ and $y = -2$ .... **(2 marks)**

(6) $s = ut + \dfrac{1}{2}at^2$

    a Find u when $s = 220$, $t = 10$ and $a = 4$ **b** Find a when $s = 88$, $t = 4$ and $u = 5$
                                          Give your answer to 1 decimal place (d.p.)

....

Exam-style question

(7) Write a formula to calculate the cost of a party booking.

PARTY BOOKINGS
ROOM HIRE **£85**
BUFFET **£12.50** PER PERSON

.... **(2 marks)**

Problem-solve!

① A car travels for $2\frac{1}{4}$ hours at an average speed of 90 km/h.
Calculate the distance travelled. ....

② The density of oak is 0.85 g/cm³.
Find the mass in kg of this oak plank.

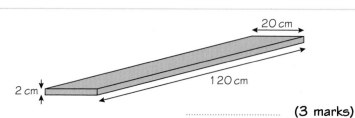

.... **(3 marks)**

③ A cyclist travels 48 km at an average speed of 19 km/h.
Calculate the time taken, in hours and minutes. .... **(3 marks)**

④ $s = ut + \frac{1}{2}at^2$ where s = distance, u = initial velocity, t = time and a = acceleration.

 a A car starts from rest and accelerates for 6 seconds.   **Hint** 'Starts from rest' means $u = 0$.
It travels 500 m.
Find its acceleration in m/s², to 1 d.p.   ....

 b Another car starts from rest and accelerates at 12 m/s².
How long does it take to travel 500 m?   .... seconds

⑤ In the formula $v^2 = u^2 + 2as$

 v is final velocity, u is initial velocity, a is acceleration and s is distance.
A car starts from rest and accelerates at 3 m/s² until
it reaches a speed of 72 km/h.   **Hint** When acceleration is in m/s²,
Calculate the distance it travels. ....   use speed in m/s in the formula.

⑥ Limousine hire costs £125 plus £17.50 per person.
Alice has £240 to pay for transport to her school prom.
For how many people can she hire the limousine?   .... **(3 marks)**

⑦ The diagram shows the dimensions of a standard DVD case.
A library needs shelves for its DVDs.

Shelves come in two lengths: 68 cm and 120 cm.
68 cm shelves cost £19.75 and 120 cm shelves cost £28.50
Work out the cheapest shelving cost for 300 DVDs.
You must show your working.

....

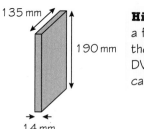

Hint Write
a formula for
the number of
DVDs a shelf
can hold.

Now that you have completed this unit, how confident do you feel?

| Finding a value that is not the subject of a formula | Deriving a simple formula | Changing the subject of a formula | Changing the subject of a more complex formula |

④ Equations

This unit will help you solve linear and quadratic equations and write equations to solve problems.

A01 Fluency check

① Solve

a $4x + 5 = 11$

b $\frac{x}{4} = 5$

c $x + 1 = 3x - 5$

$x = $ ....

$x = $ ....

$x = $ ....

② Expand and simplify

a $2(x + 5)$

b $-3(x + 4)$

c $4(x - 2) + 7$

③ Factorise

a $x^2 + 2x + 1$

b $x^2 + x - 6$

c $x^2 - 9$

④ **Number sense**

Tick the possible solutions to $a \times b = 0$

○ $a = 0, b = 3$

○ $a = 7, b = 0$

○ $a = -8, b = 0$

○ $a = -2, b = 2$

○ $a = 0, b = 0$

○ $a = 0, b = -5$

Key point

↓

Solving an equation means finding the value of the 'unknown' or mystery number represented by a letter.

These **skills boosts** will help you to write and solve different types of equations.

1 Solving linear equations with brackets > **2** Solving linear equations with fractions > **3** Solving quadratic equations > **4** Writing equations to solve problems

You might have already done some work on solving equations. Before starting the first skills boost, rate your confidence in each skill.

① Solve
$3(x + 4) = 6$

② Solve
$\frac{x}{4} + 5 = 8$

③ Solve
$x^2 + x - 2 = 0$

④ Find the value of x.

How confident are you?

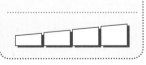 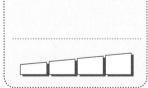 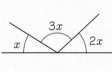

1 Solving linear equations with brackets

To solve an equations with brackets, expand the brackets first.

Worked exam question

Guided practice

Solve $3(x + 2) = 10$

$3(x + 2) = 10$

Expand the brackets.

.... $x +$ .... $= 10$

Solve the equation.

.... $x =$ ....

$x = \frac{4}{3}$

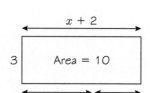

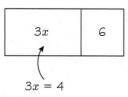

$3x = 4$

① Solve

a $2(x + 1) = 5$

b $3(x + 2) = 4x + 3$

c $2x + 11 = -3(x - 2)$

$x =$ ....

$x =$ ....

$x =$ ....

Hint Be careful multiplying negatives.

② Solve

a $2(x - 3) = 3x + 1$

b $x + 2 = -2(x + 5)$

c $4(x + 3) = 7(x - 1)$

$x =$ ....

$x =$ ....

$x =$ ....

Hint Expand all brackets first.

③ Solve

a $2(x - 5) = 5(x + 1)$

b $3(2x + 1) = -2$

c $4(2x - 3) = -3(x + 1)$

$x =$ ....

$x =$ ....

$x =$ ....

Exam-style question

Hint Expand and simplify the left-hand side.

④ Solve

$4(2x + 3) - 5 = 7$

$x =$ .... **(2 marks)**

Reflect Do equations always have positive, whole-number solutions?

2 Solving linear equations with fractions

When the letter term is a fraction like $\frac{x}{4}$ or $\frac{a}{7}$ or $\frac{2c}{3}$, first get the letter term 'on its own' on one side of the = sign, then multiply by the denominator.

Guided practice

Solve $\frac{x}{3} - 5 = 2$

 Worked exam question

Use inverse operations to get the fraction 'on its own'.

$\frac{x}{3} - 5 = 2$

$\times 3 \left(\begin{array}{c} \frac{x}{3} = \text{..........} \\ x = 21 \end{array} \right) \times 3$

Add 5 to both sides.

$x \div 3$ so use the inverse operation $\times 3$ on both sides.

$x \rightarrow \boxed{\div 3} \rightarrow \boxed{-5} \rightarrow 2$

$x \leftarrow \boxed{\times} \leftarrow \boxed{+} \leftarrow 2$

① Solve

a $\frac{x}{7} + 3 = 17$

$x = \text{..........}$

b $7 + \frac{a}{4} = 4$

$a = \text{..........}$

c $\frac{b}{2} - 5 = 6$

$b = \text{..........}$

② Solve

a $\frac{3w}{2} - 8 = 1$

$w = \text{..........}$

b $4 - \frac{2x}{7} = 5$

$x = \text{..........}$

c $2 + \frac{4x}{5} = 3$

$x = \text{..........}$

③ Simplify   **Hint** $\frac{3}{3} \times (4x + 1)$

a $\left(\frac{4x + 1}{3}\right) \times 3$

....

b $\left(\frac{3x - 5}{2}\right) \times 2$

....

c $\left(\frac{5d + 1}{7}\right) \times 7$

....

④ Solve   **Hint** $(5x + 1) \div 3$, so multiply both sides by 3.

a $x + 2 = \frac{5x + 1}{3}$

....

b $4x - 1 = \frac{2x + 3}{5}$

....

c $\frac{8 - 3t}{2} = 2t + 5$

....

Exam-style question

⑤ Solve

$\frac{2x - 1}{4} = 5 - 6x$

$x = \text{..........}$ **(3 marks)**

Reflect   Is $3 \times x + 2$ the same as $3(x + 2)$?

In Q4a, why do you need to put brackets around $(x + 2)$?

3 Solving quadratic equations

To solve a quadratic equation like $x^2 + 6x + 8 = 0$, factorise.

Guided practice

Worked exam question

Solve $x^2 + 2x - 8 = 0$

Factorise.

$$x^2 + 2x - 8 = 0$$

$(x + \text{.........})(x - \text{.........}) = 0$

So $x + \text{.........} = 0$ or $x - \text{.........} = 0$

$x = -4$ or $x = 2$

Find two numbers which multiply to give -8 and add to give 2.

$a \times b = 0$ means that either $a = 0$ or $b = 0$.

When $x = -4$, $x^2 + 2x - 8 = 16 - 8 - 8 = 0$.
When $x = 2$, $x^2 + 2x - 8 = 4 + 4 - 8 = 0$.

(1) Solve

a $x^2 - x - 6 = 0$

b $x^2 - 9x + 18 = 0$

c $x^2 + 7x + 10 = 0$

$x = \text{.........................}$

$x = \text{.........................}$

$x = \text{.........................}$

(2) Solve

a $x^2 - 4 = 0$

b $x^2 + 6x + 9 = 0$

c $x^2 - 8x + 16 = 0$

$x = \text{.........................}$

$x = \text{.........................}$

$x = \text{.........................}$

Exam-style question

(3) Solve $m^2 + 3m - 4 = 0$

$\text{.........................}$ (2 marks)

(4) Solve

a $x^2 - 5x = 0$

b $x^2 + 7x = 0$

c $2x^2 - 3x = 0$

$x = \text{.........................}$

$x = \text{.........................}$

$x = \text{.........................}$

Hint Factorise to $x(\text{.........} + \text{.........}) = 0$. One solution is $x = 0$.

(5) Solve

a $x^2 + x = 20$

b $x^2 + 2x = -8$

c $x^2 = 7x - 12$

$x = \text{.........................}$

$x = \text{.........................}$

$x = \text{.........................}$

Hint Get all the terms on one side.

Reflect How many solutions does a quadratic equation have – 0, 1, 2, more?

Writing equations to solve problems

To find an unknown number, write an equation.
Use x for the unknown number.

Guided practice

Find the missing angles in this triangle.

Worked exam question

Write an equation for the angles in this triangle.

.... + ... + ... = 180   Angles in a triangle add to 180°.

............$x + 90 = 180$ Collect like terms.

Solve to find the value of x.

............$x = $............

$x = $............

Angle $x = 30°$; angle $2x = 60°$ Remember to answer the question: write down the sizes of *both angles*.

① Find the missing angles in these triangles.

a

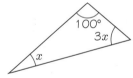

b

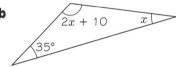

② The perimeter of this pentagon is 22 cm. Find the lengths of the sides.

③ The perimeter of this rectangle is 28 cm. Find x.

④ Find the missing angles in these quadrilaterals.

a

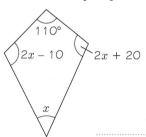

b

Hint Which angles are equal?

(5) Find the unknown angles.

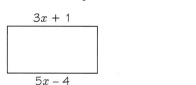

Hint Label this unknown angle x.   **Hint** Explain why this angle is x too.

....

(6) Find the length of the labelled side in this rectangle.

Hint What do you know about opposite sides of a rectangle?

3x + 1

5x – 4   ....

(7) Find the size of each angle in this parallelogram.

Hint What do you know about opposite angles in a parallelogram?

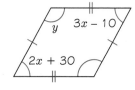

y 3x – 10

2x + 30

....

(8) The length of a rectangular garden is three times its width.
The perimeter of the garden is 56 m.
Find the width.   ....

Hint Draw a diagram.
You want to
find this.
Label it x.

Width ....

Length ....

Length = 3 × width
        = 3 × ....

(9) Sara buys a newspaper for £2.40 and two identical bottles of water.
She pays £4.20.
What is the price of one bottle of water?

Hint Use x for the unknown price.

....

(10) Tim buys 3 identical cakes and a juice, which costs £1.95.
He pays with a £10 note and gets £2.95 change.
How much does one cake cost?   ....

Exam-style question

(11) The sum of three consecutive numbers is 48.   **Hint** Use x, x + 1 and x + 2 for the three numbers.
Find the numbers.

.... **(3 marks)**

Reflect   What kind of question can you solve by writing an equation?

Practise the methods

Answer this question to check where to start.

Check up

What is the correct first step for solving this equation?

$\frac{3x}{4} + 5 = 11$

A ⃝ Multiply both sides by 4

B ⃝ Multiply both sides by 3

C ⃝ Subtract 5 from both sides

If you ticked C, finish solving the equation. Then go to Q2.

If you ticked A or B, go to Q1 for more practice.

1 Solve

a $\frac{x}{5} = 4$

b $\frac{x}{2} + 3 = 6$

c $\frac{2x}{3} - 7 = 1$

$x =$ ....

$x =$ ....

$x =$ ....

2 Solve

a $\frac{3x}{4} + 2 = 2x - 1$

b $\frac{5x}{6} - 3 = x - 6$

c $4(x + 7) = 36$

3 Solve

a $5(x - 3) = 2x + 3$

b $-5(x - 3) = 7 - 8x$

c $2x + 3 = -4(x + 7)$

4 Solve $4(2x + 3) - 8 = 3x - 5(x + 1)$

5 Solve

a $x^2 - 36 = 0$

b $x^2 - 8x = 0$

c $2x^2 + 5x = 0$

Exam-style question

6 Solve $x^2 + 5x - 14 = 0$

.... (2 marks)

Problem-solve!

 (1) Find the sizes of these angles.

....

Exam-style question

(2) ABC is an isosceles triangle with perimeter 23 cm.

Find the lengths of the unlabelled sides. .... **(3 marks)**

 (3) Multiplying a number by 4 gives the same answer as adding 18 to it.

What is the number? ....

 (4) At a theme park, adult tickets cost £24 each.
Imran buys 2 adult and 3 child tickets for £96.

How much is a child's ticket? ....

Exam-style question

(5) These two rectangles have the same area.
Find the value of x.

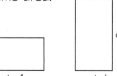

.... **(3 marks)**

 (6) I think of a number, add 5 and then divide by 4. My answer is 3.

What number did I think of? ....

 (7) Write a quadratic equation that has solutions $x = 3$ and $x = -5$.

....

Now that you have completed this unit, how confident do you feel?

1	2	3	4
Solving linear equations with brackets	Solving linear equations with fractions	Solving quadratic equations	Writing equations to solve problems

⑤ Simultaneous equations

This unit will help you to set up and solve different types of simultaneous equations.

AO1 Fluency check

① Simplify

  a $-2x + 3x$ **b** $4y - y$ **c** $3x - x$ **d** $-2y + (-2y)$

  ....

② Solve

  a $3x = 9$ **b** $2y = -8$ **c** $x + 3 = 10$ **d** $3 + 2y = -4$

  $x = $ .... $y = $ .... $x = $ .... $y = $ ....

③ Substitute $y = 2$ into $3x + 2y = 16$ and solve to find the value of x.

  $x = $ ....

④ **Number sense**

Use the operations to complete each equation.

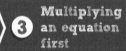 $x + y = 10$ 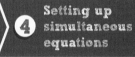

                     $\downarrow \times 5$

$2x + 2y = $ .... $x + $ .... $y = 50$ $3x + 3y = $ ....

Key points

| Simultaneous equations are two (or more) equations that involve two (or more) letters. | Solving simultaneous equations means finding the value of all the letters. |

These **skills boosts** will help you to write and solve simultaneous equations.

> **1** Subtracting to eliminate a variable   **2** Adding to eliminate a variable   **3** Multiplying an equation first   **4** Setting up simultaneous equations

You might have already done some work on simultaneous equations. Before starting the first skills boost, rate your confidence with these questions.

①
$x + 2y = 7$
$x + y = 4$

②
$3x + 5y = 32$
$x - 5y = -16$

③
$5x + 2y = 12$
$4x + y = 9$

④
3 shirts and 2 hats cost £61.
2 shirts and 2 hats cost £46.
How much is a shirt?
How much is a hat?

How confident are you?

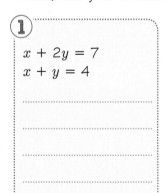 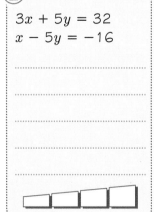 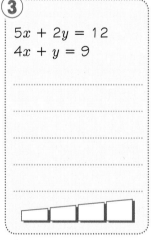 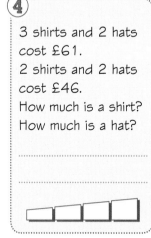

1 Subtracting to eliminate a variable

To solve a pair of simultaneous equations, add or subtract them to eliminate one of the variables.
When there are two identical terms with the same sign, subtract one equation from the other.

Guided practice

Worked exam question

Solve
$$2x + 3y = 9$$
$$2x + y = 7$$

___

Label one equation A and one equation B.
$2x + 3y = 9$ (A)
$2x + y = 7$ (B)

Subtract B from A.
$3y - y = $ .... (A) − (B)

Solve the equation to find the value of y.
$y = $ ....

Substitute $y = $ .... into A.
$2x + $ .... $ = 9$
$2x = $ ....
$x = 3$
Solution $x = 3$, $y = 1$

$2y = 2$ so $y = 1$

(A) | x | x | y | y | y |

0   7   9

(B) | x | x | y |

6   1
$x = 3$

Check your solutions work
for equation B.
$2 \times 3 + 1 = 7$ ✓

1 a Solve
$$2x + 4y = 14$$
$$2x + y = 5$$

$x = $ .... , $y = $ ....

b Solve
$$x + 3y = 10$$
$$x - y = 2$$

$x = $ .... , $y = $ ....

c Solve
$$x + 5y = 7$$
$$3x + 5y = 11$$

$x = $ .... , $y = $ ....

2 a Solve
$$4x - 6y = -18$$
$$4x + 2y = 22$$

$x = $ .... , $y = $ ....

b Solve
$$5x + 2y = 17$$
$$6x + 2y = 20$$

$x = $ .... , $y = $ ....

c Solve
$$2x + 3y = 5$$
$$2x + 4y = 8$$

$x = $ .... , $y = $ ....

Hint Simultaneous equations can have negative solutions.

Exam-style question

3 Solve
$$x + 3y = 11$$
$$2x + 3y = 16$$

$x = $ .... , $y = $ .... **(3 marks)**

4 a $3x - 2y = 10$
$x - 2y = 2$

$x = $ .... , $y = $ ....

b $2x - y = 8$
$5x - y = 23$

$x = $ .... , $y = $ ....

c $2x + y = 13$
$5x + y = 4$

$x = $ .... , $y = $ ....

Reflect How do you decide which equation to subtract from the other?

2 Adding to eliminate a variable

When two terms have the same coefficient and variable but different signs, add the equations to eliminate the variable.

Guided practice

Solve
$2x - y = 11$
$x + y = 10$

Label one equation A and one equation B.
$2x - y = 11$ Ⓐ
$x + y = 10$ Ⓑ

Add A and B and solve to find x.
$3x + 0 = \text{..........}$ Ⓐ + Ⓑ
$3x = \text{..........}$
$x = \text{..........}$

Substitute $x = 7$ into B.
$\text{..........} + y = 10$
$y = \text{..........}$
Solution $x = 7$, $y = 3$

Subtracting does not eliminate a variable:
Ⓐ $\qquad 2x - y = 11$
Ⓑ $\qquad x + y = 10$
Ⓐ − Ⓑ: $\quad x - 2y = 1$
You still have x and y.
The coefficients of y are the same and the signs are different, you can add the equations to eliminate the y variable.

Check your solution works
for equation A.
$2 \times 7 - 3 = 11$ ✓

① **a** Solve
$3x - y = 5$
$x + y = 7$

$x = \text{..........}$, $y = \text{..........}$

b Solve
$x + 2y = 7$
$3x - 2y = 13$

$x = \text{..........}$, $y = \text{..........}$

c Solve
$2x + 3y = 26$
$x - 3y = 4$

$x = \text{..........}$, $y = \text{..........}$

Exam-style questions

② Solve
$-x + 3y = 1$
$x - y = 3$

$x = \text{..........}$, $y = \text{..........}$ (3 marks)

③ Solve
$x + 2y = 12$
$3x - 2y = 12$

$x = \text{..........}$, $y = \text{..........}$ (3 marks)

④ Solve
a $2x - 3y = -15$
$4x + 3y = 33$

$x = \text{..........}$, $y = \text{..........}$

b $-4x + y = -9$
$4x - 3y = -5$

$x = \text{..........}$, $y = \text{..........}$

c $-x + 3y = 1$
$x = 3 + y$

$x = \text{..........}$, $y = \text{..........}$

Hint Rearrange the second equation so that x and y are on the same side of the equals sign.

Reflect How do you decide whether to add or subtract the equations?

3 Multiplying an equation first

If the x or y coefficients are not equal, multiply one or both of the equations so either the two x terms or the two y terms have the same coefficient.

Guided practice

Worked exam question

Solve
$2x + y = -1$
$x + 4y = 10$

Label one equation A and one equation B.
$$2x + y = -1 \quad \text{Ⓐ}$$
$$x + 4y = 10 \quad \text{Ⓑ}$$

Multiply A by 4.
$$\ldots\ldots x + \ldots\ldots y = -4 \quad \text{Ⓐ} \times 4$$

The coefficient of y in equation A is 1.
$1 \times 4 = 4$, so multiplying A by 4 makes the y coefficients in A and B equal.

Subtract B from 4 × A and solve to find x.
$$8x + 4y = -4 \quad 4 \times \text{Ⓐ}$$
$$x + 4y = 10 \qquad \text{Ⓑ} \quad -$$
$$7x + 0 = -14$$
$$x = -2$$

You could multiply B by 2 instead so that both equations have the same x coefficient.

Substitute $x = -2$ into A.
$(2 \times -2) + y = -1$
$\ldots\ldots + y = -1$
$y = \ldots\ldots$
Solution $x = -2$, $y = 3$

Check your solutions work for equation B.
$-2 + (4 \times 3) = 10$ ✓

① Solve
$3x + 2y = 17$ Ⓐ
$x + y = 7$ Ⓑ

Hint Either multiply B by 3 (to make the x coefficients equal) or by 2 (to make the y coefficients equal).

$x = \ldots\ldots$, $y = \ldots\ldots$

② Solve

a $3x - y = 10$
$x + 2y = 8$

b $x + 2y = -1$
$-3x + y = 10$

c $2x - 3y = -4$
$x + y = -7$

$x = \ldots\ldots$, $y = \ldots\ldots$ $x = \ldots\ldots$, $y = \ldots\ldots$ $x = \ldots\ldots$, $y = \ldots\ldots$

③ Solve

a $x + 2y = 0$
$3x - y = -14$

b $5x - y = 37$
$x - 3y = 13$

c $2x + -4y = 0$
$x + 2y = 12$

$x = \ldots\ldots$, $y = \ldots\ldots$ $x = \ldots\ldots$, $y = \ldots\ldots$ $x = \ldots\ldots$, $y = \ldots\ldots$

Hint Is it easier to multiply the first equation by 3 or the second equation by 2?

4 Solve

a $10x - 5y = -55$
$x - 2y = -19$

b $2x + 5y = 1$
$-x - y = -5$

c $3x + 2y = 36$
$2x - y = 17$

$x = \text{...........} , y = \text{...........}$ $x = \text{...........} , y = \text{...........}$ $x = \text{...........} , y = \text{...........}$

5 Solve
$2x + 3y = 7$ Ⓐ
$5x + 2y = 1$ Ⓑ

Hint Multiply both equations so two terms have the same coefficient.
You could multiply A by 5 and B by 2.

$x = \text{...........} , y = \text{...........}$

6 Solve

a $4x + 3y = -5$
$3x + 5y = -1$

b $3x + 7y = -4$
$2x + 5y = -2$

c $6x + 2y = 10$
$5x + 10y = 100$

$x = \text{...........} , y = \text{...........}$ $x = \text{...........} , y = \text{...........}$ $x = \text{...........} , y = \text{...........}$

7 Solve

a $x + 2y = 3$
$4x + 2y = 9$

b $5x - y = 5$
$10x + 2y = -2$

c $x + 5y = -1$
$8x - 20y = 28$

$x = \text{...........} , y = \text{...........}$ $x = \text{...........} , y = \text{...........}$ $x = \text{...........} , y = \text{...........}$

Hint Solutions to simultaneous equations can be fractions.

8 Solve

a $8x + 2y = 2$
$-4x - 3y = 3$

b $2x - 5y = -14$
$-3x + 2y = -1$

c $-4x + 3y = 13$
$6x - 7y = -32$

$x = \text{...........} , y = \text{...........}$ $x = \text{...........} , y = \text{...........}$ $x = \text{...........} , y = \text{...........}$

Exam-style question

9 Solve
$4x + 2y = 5$
$8x - 3y = 24$

$x = \text{...........} , y = \text{...........}$ **(3 marks)**

Reflect   How did you decide which equation to multiply?

4 Setting up simultaneous equations

Guided practice

2 cookies and 3 sandwiches cost £8.
5 cookies and 1 sandwich cost £7.
Find the cost of **a** a cookie **b** a sandwich.

Write the equations using x and y.

Let x = cost of a cookie
and y = cost of a sandwich.

Label the equations A and B.

$2x + 3y = 8$ Ⓐ
$5x + y = 7$ Ⓑ

Multiply B by 3.

$15x + \text{.........} y = \text{.........}$ $3 \times$ Ⓑ

Subtract A from $3 \times$ B and solve to find x.

$$15x + 3y = 21 \quad 3 \times Ⓑ$$
$$2x + 3y = 8 \quad Ⓐ \quad -$$
$$13x + 0 = \text{.........}$$
$$x = \text{.........}$$

1 cookie costs £x
2 cookies cost £$2x$

Substitute $x = 1$ into equation A and solve to find y.

a A cookie costs £...........   **b** A sandwich costs £...........

£8

Ⓐ | x | x | y | y | y |

2   3
cookies   sandwiches

£7

Ⓑ | x | x | x | x | x | y |

5   1
cookies   sandwich

① At a café 3 lemonades and 1 cola cost £5. 2 lemonades and 4 colas cost £10. Work out the cost of

a a lemonade ....   **b** a cola. ....

② 4 juices and 1 cake cost £3. 2 juices and 3 cakes cost £4. Work out the cost of

a a juice ....   **b** a cake. ....

③ Tickets for 2 adults and 3 students cost £38. The cost for 5 adults and 2 students is £62. Work out the price of

a an adult ticket ....   **b** a student ticket. ....

④ Entry for 2 adults and 5 children costs £41. Entry for 3 adults and 2 children costs £34. Work out the price of

a adult entry ....   **b** child entry. ....

Exam-style question

⑤ 4 pens and 3 notebooks cost £17. 6 pens and 2 notebooks cost £15.50. Find the cost of

a a pen ....   **b** a notebook. .... (5 marks)

Reflect Do you always need to use x and y when setting up equations?

Practise the methods

Answer this question to check where to start.

Check up

What is the simplest way to start solving these equations?

a $x + 5y = 11$

b $5x + y = 7$

 A ◯

Multiply **a** by 7

 B ◯

Multiply **b** by 5

C ◯

Multiply **a** by 5

D ◯

Subtract **b** from **a**

If you ticked B or C finish solving the equations. Then go to Q2.

If you ticked A or D go to Q1 for more practice.

1 Solve

a $2x + y = 32$
$3x + y = 42$

$x = $ .... , $y = $ ....

b $x + 2y = 9$
$3x - 2y = 11$

$x = $ .... , $y = $ ....

c $5x + 2y = 21$
$-5x + y = 3$

$x = $ .... , $y = $ ....

2 Solve

a $2x + y = 10$
$x - 2y = -5$

$x = $ .... , $y = $ ....

b $3x - 2y = 18$
$x + 4y = 13$

$x = $ .... , $y = $ ....

c $24x + 3y = 0$
$2x + y = -2$

$x = $ .... , $y = $ ....

3 Solve

a $2x + 4y = 0$
$3x + 2y = 12$

$x = $ .... , $y = $ ....

b $x + 2y = 0$
$3x - y = -14$

$x = $ .... , $y = $ ....

c $8x - y = 7$
$2x - 3y = 10$

$x = $ .... , $y = $ ....

4 Solve

a $3x + 7y = 15$
$5x + 2y = -4$

$x = $ .... , $y = $ ....

b $3x + 6y = 3$
$4x - 4y = -5$

$x = $ .... , $y = $ ....

c $3x + 5y = 3$
$12x + 10y = -2$

$x = $ .... , $y = $ ....

5 At a fair the cost of a ride for 1 adult and 3 children is £11.
The cost for 2 adults and 2 children is also £11.
Work out the cost of

a an adult ticket ....

b a child ticket. ....

Exam-style question

6 Solve
$2x + 5y = -19$
$3x - 4y = 29$

$x = $ .... , $y = $ .... **(3 marks)**

Problem-solve!

1 At the cinema tickets for 2 adults and 3 children cost £33.
Tickets for 1 adult and 1 child cost £14.
Write and solve a pair of simultaneous equations to find the cost of

a an adult ticket ....

b a child ticket. ....

2 x and y are two numbers.
Their sum $x + y$ equals 16.
Their difference $x - y$ equals 6.
What are the values of x and y? $x = $ .... , $y = $ ....

3 Find the two numbers whose difference is 15 and whose sum is 35.   .... , ...

Exam-style questions

4 Solve these simultaneous questions

$4x = y + 7$
$10x - 2y = 15$ $x = $ .... , $y = $ .... **(4 marks)**

5 Jane's wages are calculated using the formula
Wages = number of hours, n × hourly rate, h + bonus, b
$W = nh + b$
When she works 30 hours her wage is £330.
When she works 34 hours her wage is £370.
What is Jane's hourly rate and what is her bonus? $h = $ .... , $b = $ .... **(3 marks)**

---

6 Write equations and solve to find the cost of these drinks and snacks.

a 3 teas and 1 coffee cost £7.
2 teas and 4 coffees cost £13.   ....

b 4 milkshakes and 1 cookie cost £8.20.
2 milkshakes and 3 cookies cost £7.10.   ....

7 Three first class stamps and one second class stamp cost £2.47.
One first class stamp and two second class stamps cost £1.74.
What is the price difference between first and second class stamps?   ....

8 Why is it not possible to solve these simultaneous equations?
A $3x - 8y = 10$
B $6x - 16y = 20$ ....

Now that you have completed this unit, how confident do you feel?

1 Subtracting to eliminate a variable

2 Adding to eliminate a variable

3 Multiplying an equation first

4 Setting up simultaneous equations

⑥ Indices

This unit will help you to simplify expressions with indices.

AO1 Fluency check

① Work out

a 5^2 **b** 2^3 **c** $(-4)^2$ **d** $(-1)^3$

② Simplify

a $\dfrac{1}{-2}$ **b** $\dfrac{5}{-7}$ **c** $\dfrac{4}{1}$ **d** $\dfrac{6}{-1}$

③ Calculate

a $\dfrac{2}{3} \times \dfrac{4}{5}$ **b** $\dfrac{1}{3} \times \dfrac{3}{8}$ **c** $1 \div \dfrac{1}{5}$ **d** $1 \div \dfrac{2}{7}$

④ **Number sense**

Simplify

a $\dfrac{2 \times 2 \times 2}{2}$ **b** $\dfrac{5 \times 5}{5 \times 5 \times 5}$ **c** $\dfrac{4 \times 4 \times 4 \times 4}{4 \times 4}$

Key points

A positive index number (or power) tells you how many times a number is multiplied by itself.
$3^4 = 3 \times 3 \times 3 \times 3$

A negative index tells you to take the reciprocal. The reciprocal of a number is 1 divided by the number.

These **skills boosts** will help you simplify and calculate with indices.

| ① Laws of indices for multiplication and division | ② Reciprocals | ③ Zero and negative indices |

You might have already done some work with indices. Before starting the first skills boost, rate your confidence using each skill.

① Simplify $3^2 \times 3^3$

② Find the reciprocal of $\dfrac{1}{4}$

③ Evaluate 2^{-3}

How confident are you?

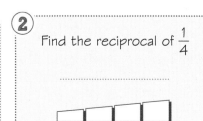

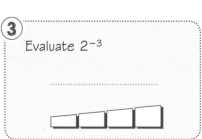

Laws of indices for multiplication and division

To multiply two powers of the same number, add the powers.
To divide two powers of the same number, subtract the powers.

Guided practice

a Simplify $5^3 \times 5^4$

b Simplify $\dfrac{4^4}{4^3}$

Write as a single power.

a $5^3 \times 5^4 = 5^{3\,+\,......}$

$\qquad = 5^{......}$

$\underbrace{5 \times 5 \times 5}_{5^3} \times \underbrace{5 \times 5 \times 5 \times 5}_{5^4} = 5 \times 5 \times 5 \times 5 \times 5 \times 5 \times 5 = 5^7$

b $\dfrac{4^4}{4^3} = 4^{4\,-\,......}$

$\qquad = \text{..........}$

$\dfrac{4^4}{4^3} = \dfrac{\cancel{4} \times \cancel{4} \times \cancel{4} \times 4}{\cancel{4} \times \cancel{4} \times \cancel{4}} = \dfrac{4^1}{1} = 4$

① Simplify

a $3^2 \times 3^5$

b $4^5 \times 4$ **Hint** $4^1 = 4$

c $2^3 \times 2 \times 2^5$

② Write as a single power.

a $\dfrac{2^7}{2^3}$

b $\dfrac{10^5}{10}$

c $5^6 \div 5^4$

③ Evaluate

a $2^2 \times 2^3$

b $\dfrac{3^5}{3^3}$

c $\dfrac{10^9}{10^6}$

Hint Simplify to $\square^{\square}$, then work out the value.

④ Simplify

a $\dfrac{5^3 \times 5^2}{5}$

b $\dfrac{7^3 \times 7}{7^2}$

c $\dfrac{4^2 \times 4^5}{4^4}$

Exam-style question

Hint $\dfrac{(4^2)^3}{4^2 \times 4^2 \times 4^2}$

⑤ Write $(4^2)^3$ as a single power.

(1 mark)

Reflect What do 'evaluate' and 'simplify' mean in indices questions?

2 Reciprocals

The reciprocal of a number n is: $1 \div n$ or $\dfrac{1}{n}$.

Guided practice

Find the reciprocal of

a 5   **b** $\dfrac{1}{3}$

Work out 1 divided by the number.

a $1 \div \underline{\quad\quad} = \dfrac{1}{5}$

b $1 \div \dfrac{1}{3}$

$= 1 \times \dfrac{\underline{\quad\quad}}{1}$

$= \dfrac{\underline{\quad\quad}}{1}$

$= 3$

$1 \div 5 = \dfrac{1}{5}$

Dividing by $\dfrac{1}{n}$ is the same as multiplying by $\dfrac{n}{1}$

1 can be written as $\dfrac{1}{1}$. Multiply the numerators then multiply the denominators.

---

(1) Find the reciprocal of

a 6

b $\dfrac{1}{8}$

c $-\dfrac{1}{2}$

....

(2) Find the reciprocal of

a $\dfrac{3}{5}$

b $-\dfrac{5}{7}$

c $1\dfrac{1}{2}$

Hint Change to an improper fraction first.

....

(3) Complete the inputs and outputs of this function machine.

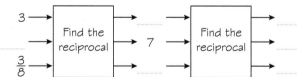

(4) **a** Find the reciprocal of 12.

b Work out 12 × reciprocal of 12.

....

(5) **a** Find the reciprocal of $\dfrac{4}{9}$.

b Work out $\dfrac{4}{9}$ × reciprocal of $\dfrac{4}{9}$.

....

Exam-style question

(6) Find the reciprocal of 0.2.   **Hint** $1 \div 0.2$

.... **(2 marks)**

Reflect

What happens when you multiply a number by its reciprocal?
What is the reciprocal of the reciprocal of a number?

3 Zero and negative indices

Any number to the power 0 is equal to 1.
Any number to the power −1 is the reciprocal of the number.

Guided practice

Worked exam question

Evaluate

a 8^0 **b** 3^{-1} **c** $\left(\dfrac{5}{7}\right)^{-1}$

a $8^0 = 1$

$8 \times 1 = 8$ and $8 \times 8^0 = 8^{1+0} = 8$. So $8^0 = 1$.

b $3^{-1} = 1 \div \text{............}$

$\quad = \dfrac{1}{3}$

3^{-1} is the reciprocal of 3.

$3^0 \div 3 = 3^{0-1} = 3^{-1}$ and $3^0 \div 3 = 1 \div 3 = \dfrac{1}{3}$

So $3^{-1} = \dfrac{1}{3}$

c $\left(\dfrac{5}{7}\right)^{-1} = 1 \div \dfrac{5}{7}$

$\quad = 1 \times \dfrac{7}{\text{............}}$

$\quad = \dfrac{7}{5}$

$\left(\dfrac{5}{7}\right)^{-1}$ is the reciprocal of $\dfrac{5}{7}$

(1) Evaluate

a 12^0 **b** 6^{-1} **c** $\left(\dfrac{3}{4}\right)^{-1}$ **d** $\left(2\dfrac{1}{2}\right)^{-1}$

(2) Write as a single power.   **Hint** Write as an improper fraction first.

a $\dfrac{2^5}{2^2}$ **b** $4^3 \times 4^0$ **c** $5^3 \times 5^{-1}$ **d** $\dfrac{3^2}{3^5}$

(3) Work out the value of

Hint To find a negative power deal with the negative (find the reciprocal) first.
$3^{-2} = \left(\dfrac{1}{3}\right)^2 = \dfrac{1}{3^2}$

a 3^{-2} **b** 5^{-2} **c** $\left(\dfrac{1}{2}\right)^{-2}$

Exam-style question

(4) Evaluate

a 2^{-2} .... (1 mark)   **b** $(-10)^{-3}$ .... (1 mark)   **c** $\left(\dfrac{1}{3}\right)^{-3}$ .... (1 mark)

Reflect   Is the reciprocal of a whole number greater than or less than 1?

Practise the methods

Answer this question to check where to start.

Check up

Tick the correct simplification of $\dfrac{2^3 \times 2}{2^5}$

 A 2^{-3} ◯

 B $2^{-1} = \dfrac{1}{2}$ ◯

 C 2 ◯

D $2\dfrac{4}{5}$ ◯

If you ticked B go to Q3.

If you ticked A, C or D go to Q1 for more practice.

1 Simplify

a $5^2 \times 5 = 5^{\ldots + \ldots} = \ldots$

b 7×7^5

c $\dfrac{4^3}{4}$

2 Write as a single power.

a $\dfrac{3^5}{3^6} = 3^{\ldots - \ldots} = \ldots$

b $\dfrac{10}{10^3}$

c $\dfrac{5^2}{5^2}$

3 Write as a single power.

a $(3^5)^2$

b $(2^3)^6$

c $(7^3)^4$

Exam-style question

4 a Evaluate 4^3 .... (1 mark)

b Write down the reciprocal of $\dfrac{3}{5}$ .... (1 mark)

c Work out the value of $\left(\dfrac{1}{5}\right)^{-2}$ .... (1 mark)

5 Work out the value of

a $(-1)^2 \times (-1)^3$

b $\dfrac{6^3 \times 6^2}{6^4}$

c $\dfrac{7^2 \times 7^3}{7^5}$

6 Evaluate

a 7^0

b 8^{-2}

c $(0.5)^{-1}$

d 2^{-4}

e $(-3)^{-2}$

f $\left(\dfrac{1}{2}\right)^{-3}$

Problem-solve!

① Show that $\frac{5}{9}$ multiplied by its reciprocal equals 1.

....

② Simplify

a $x^2 \times x^5$ **b** $y \times y^4$ **c** $\dfrac{z^3}{z}$

....

③ Match the equivalent expressions. One has been done for you.

x^{-1} x^{-2}

$\dfrac{1}{2x}$ 1

$\dfrac{1}{x^2}$ $\dfrac{1}{x}$

$\dfrac{2}{x}$ $(2x)^{-1}$

x^0 $\left(\dfrac{x}{2}\right)^{-1}$

④ Simplify

a $\dfrac{a^2 \times a^4}{a^5}$ **b** $\dfrac{m}{m \times m^4}$ **c** $n^{-2} \times n^5$

....

Exam-style questions

⑤ Simplify $\dfrac{x^4 \times x^2}{x^3}$ .... **(2 marks)**

⑥ Use a calculator to work out

$8.2^2 \times 8.2^{-3}$

a Write down all the digits on your calculator display.   .... **(2 marks)**

b Round your answer to 3 significant figures.   .... **(1 mark)**

Now that you have completed this unit, how confident do you feel?

1 — Laws of indices for multiplication and division

2 — Reciprocals

3 — Zero and negative indices

(7) Pythagoras' theorem

This unit will help you calculate the lengths of sides and solve problems involving right-angled triangles.

AO1 Fluency check

(1) Work out

    a $6^2 = $ ....   **b** $5^2 + 8^2 = $ ....

(2) Find

    a $5^2 + 1.4^2 = $ ....   **b** $\sqrt{18.29} = $ ....

(3) Find $a^2 + b^2$, where $a = 7$ and $b = 4$.

(4) Solve

    a $c^2 = 49$ **b** $25 + x = 49$

(5) Which of these are right-angled triangles?

A   **B**   **C**   **D**

(6) Number sense

Round each number to 1 decimal place (d.p.) and then match the pairs. Which is the odd one out?

| 8.4719 | 8.36 | 8.194 | 8.4052 | 8.253 | 8.349 | 8.08.... | 8.5487 | 8.2171 |

Key point
↓

Pythagoras' theorem connects the lengths of the three sides in a right-angled triangle.
Label the sides of the right-angled triangle a, b, c, with a the shortest side and c the longest side.
Then Pythagoras' theorem says that $c^2 = a^2 + b^2$

These **skills boosts** will help you to solve problems involving right-angled triangles.

1 Using Pythagoras to find the longest side in a right-angled triangle

2 Using Pythagoras to find one of the shorter sides in a right-angled triangle

3 Solving problems involving right-angled triangles

You might have already done some work involving right-angled triangles. Before starting the first skills boost, rate your confidence in each use of these problems using Pythagoras.

(1) Find x.

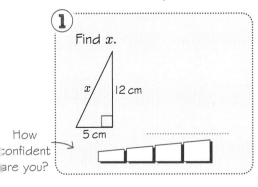

x 12 cm   5 cm

(2) Find y to 1 d.p.

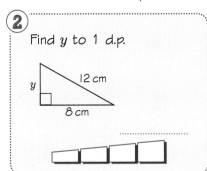

y 12 cm   8 cm

(3) A ladder leans against a wall. Its base is 1.8 m from the wall. It reaches 2.7 m up the wall. Find the length of the ladder. ....

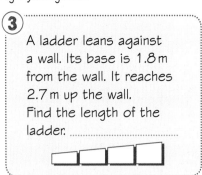

How confident are you?

Using Pythagoras to find the longest side in a right-angled triangle

Guided practice

Find x.

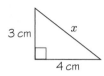
3 cm
x
4 cm

Use Pythagoras' theorem.

$c^2 = a^2 + b^2$

Substitute for a, b and c.

$x^2 = \rule{0.8cm}{0.4pt}^2 + \rule{0.8cm}{0.4pt}^2$

$ = \rule{0.8cm}{0.4pt} + \rule{0.8cm}{0.4pt}$

$ = \rule{0.8cm}{0.4pt}$

$x = \sqrt{\rule{1cm}{0.4pt}}$

$ = 5\text{ cm}$

Label the sides a, b, c.
a is the shortest side;
c is the longest.

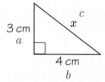
3 cm
a
c
x
4 cm
b

Take the square root of both sides and use the positive value for the length.

1 a Find y.

8 cm
6 cm
y

b Find z to 1 d.p.

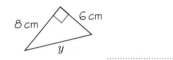
5 cm
z
7 cm

Hint The side lengths may not be whole numbers. Use a calculator.

2 a Find x to 1 d.p.

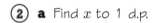
x
5 cm
3 cm

b Find y to 1 d.p.

9 cm
y
6 cm

Exam-style question

3 Find x.

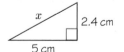
x
2.4 cm
5 cm

DIAGRAM NOT ACCURATELY DRAWN

Give your answer to 1 d.p. .... **(3 marks)**

4 a Find z to 1 d.p.

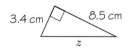
3.4 cm
8.5 cm
z

b Find c to 1 d.p.

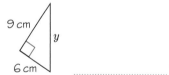
3.8 cm
c
3.8 cm

Hint a and b are the same length.

Reflect In Q3, why does it say 'DIAGRAM NOT ACCURATELY DRAWN'?

2 Using Pythagoras to find one of the shorter sides in a right-angled triangle

Guided practice

Worked exam question

Find y.

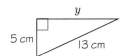

Use Pythagoras' theorem.

$$c^2 = a^2 + b^2$$

Label the sides a, b, c.

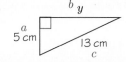

Substitute for a, b, and c.

$$\underline{}^2 = \underline{}^2 + y^2$$

$$\underline{} = \underline{} + y^2$$

Substitute the values from the diagram.

Solve to find y.

$$y^2 = \underline{} - \underline{}$$

Get the y^2 term on its own.

$$y = \sqrt{\underline{}}$$

Take the square root of both sides.

$$y = 12\,\text{cm}$$

① **a** Find x.

b Find x to 1 d.p.

c Find z to 1 d.p.

② Find the missing lengths, to 1 d.p.

a

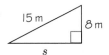

b

c

 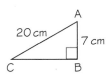

Exam-style question

③ In triangle ABC:

$\angle ABC = 90°$

AB = 7 cm

AC = 20 cm.

Find the length of CB.

Give your answer to 1 d.p.

DIAGRAM NOT ACCURATELY DRAWN

(3 marks)

Reflect   Maria says, 'If you label the longest side c, it doesn't matter which of the shorter sides is a and which is b'. Find the missing lengths in these identical triangles to see if she is correct.

3 Solving problems involving right-angled triangles

Guided practice

A ladder leans against a wall.
The base of the ladder is 2.5 m from the wall.
The ladder is 3.8 m long.
How far up the wall does the ladder reach?
Give your answer to the nearest centimetre.

$c^2 = a^2 + b^2$

$\underline{\quad}^2 = \underline{\quad}^2 + b^2$

$b^2 = \underline{\quad} - \underline{\quad}$

$b = \sqrt{\underline{}}$

wall = 2.86 m = 286 cm

A wall is at right angles to the ground.
Add the lengths you are given to the diagram.

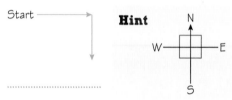

(1) A ship sails 5 km due East and then 4 km due South, then directly back to the start.

  a Mark the distances and the right angle on the diagram.

  b Calculate the distance the ship sails back to the start. ....

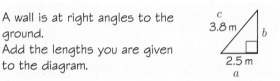

Hint

Exam-style question

(2) A helicopter flies 10 km due South from airport A.
It turns and flies 7 km due West to airport B.
Then it flies directly from airport B back to airport A.
Calculate the total distance the helicopter flies. .... **(4 marks)**

(3) The diagram shows points A (2, 4) and B (4, 10).

  a Write the length of side b on the diagram.

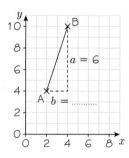

  b Use Pythagoras' theorem to find c, the length of AB. ....

(4) Calculate the length of the line joining points E (−3, 2) and F (5, 4). **Hint** Draw a diagram.

....

Reflect

What strategies did you use to solve right-angled triangled triangle problems?

Practise the methods

Answer this question to check where to start.

Check up

Tick the correct calculation to find x.

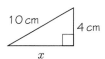
10 cm   4 cm
x

A ◯

$x^2 = 16 + 100$

$x = \sqrt{116}$

B ◯

$100 - 16 = x^2$

$x = \sqrt{84}$ cm

C ◯

$16 + x^2 = 100$

$x = 116$ cm

If you ticked B go to Q2.	If you ticked A or C, go to Q1 for more practice.

1 Solve to find x. Give your answers to 1 d.p.

  a $25 + x^2 = 81$ **b** $x^2 + 49 = 100$ **c** $64 + x^2 = 144$

      $x^2 = 81 - \text{............}$

      $x = \sqrt{\text{............}}$

      $x = \text{............................}$

2 Find x to 1 d.p.

  a
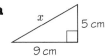
x 5 cm
9 cm

  b

5.1 cm x
2.8 cm

  c
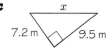
x
7.2 m   9.5 m

3 Find the unknown length in each triangle. Give your answers to 1 d.p.

  a

x 12 cm
6 cm

  b
56 mm y
11 mm

  c
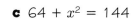
z 1.2 km
5 km

Exam-style question

4 A phone mast is supported by cables as shown in the diagram.
The phone mast is 20 m tall.
Each cable is 22 m long.
Find x, the distance between the bottom of the mast
and the point where the cable is attached to the ground.

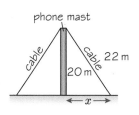
phone mast
cable   cable   22 m
20 m
← x →

                                                    (3 marks)

5 Find the length of CD, where C is the point (1, 6)
and D is the point (5, 2).

Problem-solve!

(1) A ladder 4 m long leans against the wall of a house.
The base of the ladder is 1.2 m from the wall.
Will the ladder reach a window 3.2 m up the wall?
Show your working to explain your answer. .... (3 marks)

(2) ABCD is a rectangle.
Calculate the length of
the diagonal AC, to
1 d.p.

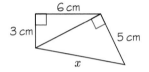

DIAGRAM NOT
ACCURATELY DRAWN

.... (3 marks)

(3) Find x.

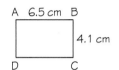

.... (4 marks)

(4) An aeroplane pilot wants to fly 40 km due East and then 50 km due South, before flying directly back to his starting position. The aeroplane has enough fuel to fly 150 km.
Is this enough for the flight?
Show your working to explain your answer. ....

(5) Does this triangle have a right angle?
Explain your answer.

Hint If the triangle does have a right angle, which angle must it be?
Does $c^2 = a^2 + b^2$?

....

(6) Two lines, AB and AC, are drawn on a 1 cm coordinate grid.

a Which line is longer? ....

b By how much?
Give your answer correct to 3 d.p. ....

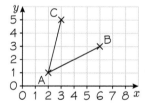

(7) ABC is an isosceles right-angled triangle.
Find the lengths of AB and BC.

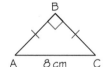

Hint Label both AB and BC x, and use Pythagoras' theorem.

....

Now that you have completed this unit, how confident do you feel?

1 Using Pythagoras to find the longest side in a right-angled triangle

2 Using Pythagoras to find one of the shorter sides in a right-angled triangle

3 Solving problems involving right-angled triangles

⑧ The equation of a straight line

This unit will help you use and find equations of straight lines.

① Work out the value of $2x + 5$ when

 a $x = 0$ .... **b** $x = 3$ .... **c** $x = -2$ ....

② $5x + 2y = 10$

 a Find y when $x = 0$ .... **b** Find x when $y = 0$ ....

③ Rearrange each equation to make y the subject.

 a $3x + y = 4$ .... **b** $7x + 2y = 11$ .... **c** $x - 5y = 9$ ....

④ **Number sense**

 a $2 \times (-2) - 1 =$ .... **b** $2 \times (-1) - 1 =$ .... **c** $2 \times 0 - 1 =$ ....

Key points

The graph of the equation $y = mx + c$ is a straight line with gradient m and y-intercept c.	The equation of a straight line can also be given in the form $ax + by = c$.	The variable 'c' can stand for any constant number in the equation of a line.

These **skills boosts** will help you find and use the equation of a straight line.

1 Finding the equation of a straight line from its graph ▷ **2** Using the equation $y = mx + c$ ▷ **3** Using the equation $ax + by = c$

You might have already done some work on equations of straight lines. Before starting the first skills boost, rate your confidence on each skill.

①
Find the equation of this line.

②
Write down the gradient and y-intercept of the line $y = 4x - 5$

③
Find the gradient and y-intercept of the line $4x + 3y = 6$

How confident are you?

1 Finding the equation of a straight line from its graph

To find the equation of a straight line, $y = mx + c$, find the gradient m and the y-intercept c.

Guided practice

Find the equation of the line.

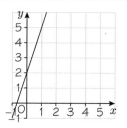

Worked
exam
question

_____

Find the y-intercept.

$c =$ ....

Find the gradient.

$m =$

Substitute m and c into $y = mx + c$

$y =$ .... $x +$ ....

Find c, where the line crosses the y-axis.

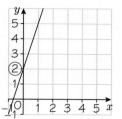

Find m, how much the graph goes up when x increases by 1.

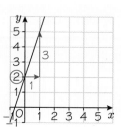

① Find the equation of each of these straight lines.

a $y =$ ....

b $y =$ ....

c $y =$ ....

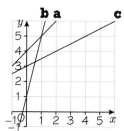

② Find the equation of each of these straight lines.

a $y =$ ....

b $y =$ ....

c $y =$ ....

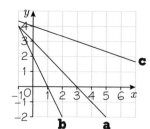

Hint A negative gradient is how much the graph goes down when x increases by 1.

Exam-style question

③ Find the equation of this line.

.... (2 marks)

Reflect Given the equations of three lines, how can you tell which line is the steepest?

2 Using the equation $y = mx + c$

Parallel lines have the same gradient, m. When two equations have the same m value (before the x) the two lines are parallel.

Guided practice

Worked exam question

Draw the graph of $y = 2x - 1$ for values of x between -2 and 2.

Make a table of values for the equation $y = 2x - 1$

x	-2	-1	0	1	2
y	-5			1	

Work out $y = 2x - 1$ when $x = -2$

Plot the points on the grid and join them with a straight line.

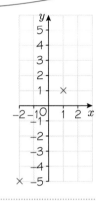

Label the line $y = 2x - 1$

① Draw the graph of $y = 2x + 3$ for values of x between -2 and 2.

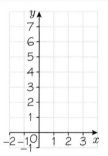

② Which of these lines are parallel?

A $y = 3x + 5$ **B** $y = 6x + 10$ **C** $y = -3x + 2$
D $y = 3 + 5x$ **E** $y = 3x$ **F** $y = 3x - 2$ ....

Hint Which equations have the same m value?

③ Write the equation of a line parallel to

a $y = 3x + 4$ ....

b $y = -5x + 1$ ....

Hint For a parallel line: gradient is the same, y-intercept is different.

④ Write the coordinates of the point where each of these lines crosses the y-axis. **Hint** $(0, \square)$

a $y = 6x + 2$ .... **b** $y = 4x - 1$ .... **c** $y = 5x$ ....

Exam-style question

⑤ Line A is parallel to $y = 4x + 3$ and intercepts the y-axis at $(0, 5)$.

Write down the equation of line A. .... **(2 marks)**

Reflect What is the same about the lines $y = 2x + 3$ and $y = 2x - 1$? What is different?

3 Using the equation $ax + by = c$

To draw a graph of an equation $ax + by = c$, make a table of values for $x = 0$ and $y = 0$.

Guided practice

Draw the graph of $x + 2y = 4$

x	O	....
y	....	O

Make a table of values for $x = 0$ and $y = 0$

Find x when $y = 0$

Find y when $x = 0$

Plot the two points on the axes and join them with a straight line.

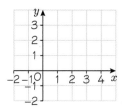

Or you could use a table of values like this.

x	-2	-1	-O	1	2
y					

Label the line $x + 2y = 4$

(1) Draw the graph of $2x + 3y = 6$

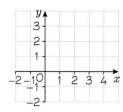

(2) Draw the graph of $2x - y = 2$ on the same grid as in Q1.

(3) **a** Rearrange $2x - y = 2$ to make y the subject. ....

    b Write down the gradient and the y-intercept. ....

      Hint Use the rearranged equation $y = mx + c$

(4) Which of these lines are parallel?   **Hint** Rearrange to $y = mx + c$

    A $-2x + y = 3$ **B** $6x + 2y = 9$ **C** $x + 2y = 3$

    D $2x + 5y = 8$ **E** $2x - y = 5$ **F** $4x - 2y = 7$ ....

Exam-style question

(5) Write down the gradient and the y-intercept of the line
$3x + 2y = 12$

.... **(2 marks)**

Reflect Why is it quicker to use $x = 0$ and $y = 0$ to find points for the graph, than to use x-values from -2 to 2? Does this work for $y = mx + c$ graphs too?

Practise the methods

Answer this question to check where to start.

Check up

What is the equation of this line?

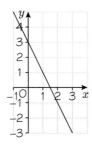

A $y = 2x + 3$

B $y = 3x + 2$

C $y = -2x + 3$

D $y = -\frac{1}{2}x + 3$

If you ticked C go to Q2.

If you ticked A, B or D go to Q1 for more practice.

(1) Find the equation of each line.

  a $y = mx + c = -1 \times x +$ .... $=$ ....

  b $y =$ .... $x + 1 =$ ....

  c $y =$ ....

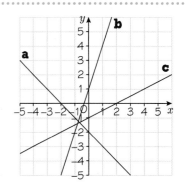

(2) **a** Draw the graph of $y = 3x - 1$ for values of x from -2 to 2.

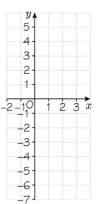

  b Draw the graph of $5x + 3y = 15$ on the same axes.

Exam-style questions

(3) Which of these lines are parallel?

  A $y = 6x + 3$ **B** $y = 6x - 2$ **C** $y = -6x + 3$ **D** $y = 3x - 2$

.... **(2 marks)**

(4) Write the equation of any line parallel to

  a $y = 4x - 1$ .... **(1 mark)**

  b $3x - 6y = 7$ .... **(2 marks)**

Problem-solve!

(1) Write these in order of steepness of the lines, starting with the least steep.

A $y = \frac{1}{2}x + 5$ **B** $y = 2x + 10$ **C** $y = \frac{3}{4}x - 2$ **D** $y = \frac{3}{2}x - 1$ ....

Exam-style questions

(2) Which of the lines **A** $y = \frac{1}{2}x + 3$ **B** $x + 2y = 6$ **C** $x - 2y = 4$

    a are parallel? .... (1 mark)

    b have the same y-intercept? .... (1 mark)

(3) Write the equation of the line parallel to $y = 3x + 4$ that passes
through $(0, -2)$. .... (2 marks)

(4) Draw the graphs of $y = -2x - 1$
and $y = \frac{1}{2}x + 2$ for values of

x from -2 to 2 on the same axes.

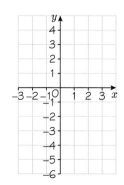

.... (6 marks)

(5) Line A has the equation $y = -2x + 4$

Write the equation of line B.

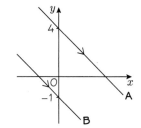

.... (2 marks)

(6) Draw graphs of $x + y = 11$
and $2x - 3y = 12$

Hence find the solution to the
simultaneous equations

    $x + y = 11$

    $2x - 3y = 12$

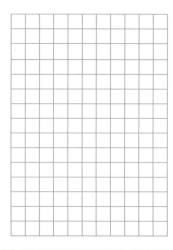

.... (4 marks)

Now that you have completed this unit, how confident do you feel?

1 Finding the equation
of a straight line
from its graph

2 Using the equation
$y = mx + c$

3 Using the equation
$ax + by = c$

⑨ Non-linear graphs

This unit will help you draw and interpret graphs that are not straight lines.

① Work out

    a 3^2 ....   **b** $(-2)^2$ ....   **c** 2^3 ....   **d** $(-1)^3$ ....

② Work out the value of $x^2 + 4$ when

    a $x = 0$ ....   **b** $x = 3$ ....   **c** $x = -1$ ....

③ Work out the value of $x^3 - 3$ when

    a $x = 1$ ....   **b** $x = 2$ ....   **c** $x = -1$ ....

④ **Number sense**

    Work out

    a $3 \times 3 + 1$ **b** $(-2) \times (-2) - 4$ **c** $(-1) \times (-1) \times (-1) + 5$

Key points

| The graph of a quadratic equation like $y = x^2 + 2x - 3$ is a U-shaped curve. | The graph of a cubic equation like $y = x^3 - 4$ is a $\curvearrowright$-shaped curve. |

These **skills boosts** will help you draw and interpret non-linear graphs.

1 Drawing quadratic graphs   **2** Finding solutions and turning points from quadratic graphs   **3** Drawing cubic graphs

You might have already done some work on non-linear graphs. Before starting the first skills boost, rate your confidence using each skill.

①
Complete this table of values for $y = x^2 + 1$

x	-2	-1	0	1	2
y					

②
Find the solutions to $x^2 - 4x + 3 = 0$

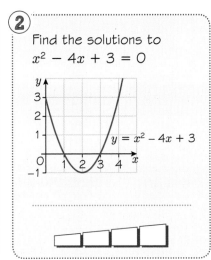

③
Draw the graph of $y = x^3 - 3x$ for values of x from -2 to 2.

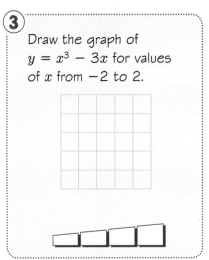

How confident are you?

1 Drawing quadratic graphs

The graph of a quadratic equation:
- with x^2 is a U-shaped curve
- with $-x^2$ is a ∩-shaped curve.

To draw a graph of an equation, first make a table of values.

Guided practice

Draw the graph of $y = x^2 - 1$ for values of x between -2 and 3.

Make a table of values. → Values of x from -2 to 3

x	-2	-1	0	1	2	3
y	3	....	....	0	....	....

Work out $x^2 - 1$ when $x = -1$

Plot the points from your table.

Join with a smooth curve.

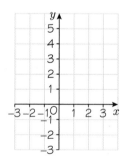

Drawn freehand
Not flat or pointed

 1 Draw these graphs for values of x between -2 and 2.

a $y = x^2$

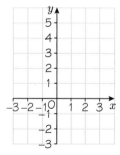

b $y = x^2 - 2$

Exam-style question

2 Draw the graph of $y = -x^2$ for values of x between -2 and 2.

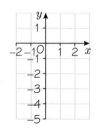

(3 marks)

Reflect A quadratic graph has line symmetry.
How can you use this symmetry to check the points you plot are correct?

2 Finding solutions and turning points from quadratic graphs

The **turning point** of a quadratic graph:

- with x^2 gives a minimum value of y.

- with $-x^2$ gives a maximum value of y.

The **solutions** of the quadratic equation $ax^2 + bx + c = 0$ are where the graph of $y = ax^2 + bx + c$ intercepts the x-axis.

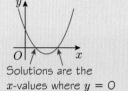

Solutions are the x-values where $y = 0$

Guided practice

Here is the graph of $y = x^2 - x - 2$

a Write down the minimum value of y.

b Find two solutions of $x^2 - x - 2 = 0$

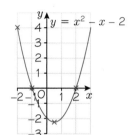

a Read the value of y at the turning point.

$y = $ ....

b Read the values of x when $y = 0$

$x = $ .... and $x = $ ....

Hint Solutions are not always whole numbers. Read the scale as accurately as you can.

---

① Use the graph of $y = x^2 - 4$

a Write down the minimum value of y.

$y = $ ....

b Find two solutions of $x^2 - 4 = 0$

$x = $ .... and $x = $ ....

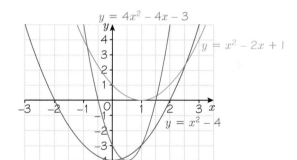

Exam-style questions

② Use the graph of $y = x^2 - 2x + 1$

a Write down the value of y at the turning point. $y = $ .... **(2 marks)**

b Find the solution of $x^2 - 2x + 1 = 0$ $x = $ .... **(1 mark)**

③ Use the graph of $y = 4x^2 - 4x - 3$

a Write down the coordinates of the turning point. .... **(1 mark)**

b Find approximate solutions to $4x^2 - 4x - 3 = 0$ .... **(2 marks)**

Reflect Look back at the graphs of quadratic equations you drew in Skills boost 1. Which of them have one solution and which have two solutions when $y = 0$?

3 Drawing cubic graphs

The graph of a cubic equation • with x^3 is this shape: • with $-x^3$ is this shape: .

a Draw the graph of $y = x^3 - 1$ for values of x between -2 and 2.

b Estimate the value of y when $x = 1.5$

a Make a table of values. → Values of x from -2 to 2

x	-2	-1	0	1	2
y	....	-2	-1	0	....

Work out $x^3 - 1$ when $x = -2$

Plot the points from your table.

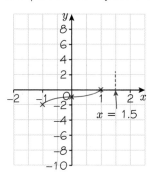

$x = 1.5$

Join with a smooth curve.

b $y =$ ....   Read off the value of y when $x = 1.5$

Use these axes for questions 1 and 2.

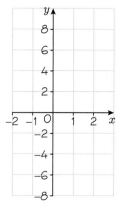

1 **a** Draw the graph of $y = x^3$ for values of x between -2 and 2.

b Estimate the value of y when $x = 1.4$ ....

2 **a** Draw the graph of $y = -x^3$ for values of x between -2 and 2.   (2 marks)

b Estimate the value of $(-1.2)^3$.   .... (1 mark)

Reflect $(-1)^3 = (-1) \times (-1) \times (-1) =$ .... and $-1^3 = -(1 \times 1 \times 1) =$ ....
Are -2^3 and $(-2)^3$ the same?

Practise the methods

Answer this question to check where to start.

Check up

Tick the correct graph for $y = -x^2 + 2$

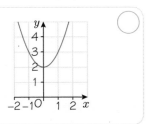

A ◯

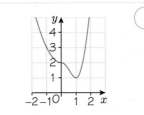

B ◯

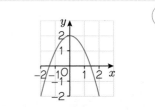

C ◯

If you ticked C, go to Q2.

If you ticked A or B, go to Q1 for more practice.

1 Draw the graph of $y = -x^2 + 1$ for values of x between -2 and 2.

x	-2	-1	0	1	2
y	....	....	....	....	....

2 Use the graph of $y = x^2 - 2x - 3$

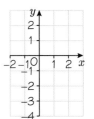

a Write down the minimum value of y.

$y =$ ....

b Find two solutions of $x^2 - 2x - 3 = 0$

$x =$ .... and $x =$ ....

Exam-style questions

3 Draw the graph of $y = x^3 + 1$ for values of x from -3 to 3.

x	-3	-2	-1	0	1	2	3
y	....	....	....	....	....	....	....

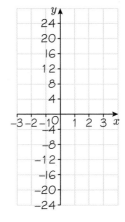

(4 marks)

4 Match each graph to its equation.

A $y = x^2 + 4$ **B** $y = x + 4$ **C** $y = x^3 + 2$ **D** $y = 3 - x^2$

i

ii

iii

iv

(3 marks)

Problem-solve!

Exam-style questions

(1) The table gives the height in metres of a rugby ball at different times (in seconds).

Time, t (seconds)	0	1	2	3	4	5
Height, h (metres)	0	4	6	6	4	0

a Plot a graph of the height of the ball against time.

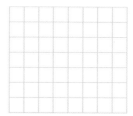

(1 mark)

b From your graph, estimate the maximum height of the rugby ball. .... (1 mark)

(2) **a** Draw the graph of $y = x^2 - 2x - 2$ for values of x between -2 and 4.

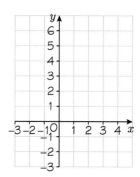

(2 marks)

b Write down the coordinates of the turning point of this graph. .... (1 mark)

c Write an equation for its line of symmetry. .... (1 mark)

(3) Here is the graph of $y = x^3 + 2$

Use the graph to estimate:

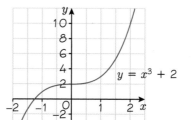

a the value of y when $x = 0.8$ .... (1 mark)

b the value of x when $y = 7$ .... (1 mark)

c the solution to $x^3 + 2 = 0$ .... (1 mark)

Now that you have completed this unit, how confident do you feel?

(1) **Drawing quadratic graphs**

(2) **Finding solutions and turning points from quadratic graphs**

(3) **Drawing cubic graphs**

⑩ Inequalities

This unit will help you solve inequalities and show inequalities on a number line.

① $-2 \leqslant x < 4$

Which values in the box

| −2.6 | 0 | −2 | 4.3 | 1.8 | −1 | 4 |

a are possible values of x?

$x = $ ....

b are possible integer values of x?

$x = $ ....

② Solve

a $\dfrac{x}{5} = 2$

$x = $ ....

b $1 = 3x + 7$

$x = $ ....

c $5x - 8 = 3x - 6$

$x = $ ....

③ Write down the inequalities shown on the number lines.

a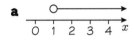

....

b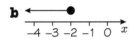

....

c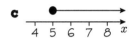

....

④ **Number sense**

Write < or > between each pair of numbers.

a 4 7   **b** −4 .... −7   **c** −4 7   **d** 4 .... −7

Key points

On a number line, ○ shows the number **is not** included.

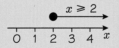

On a number line, ● shows the number **is** included.

$x \geqslant 2$

You can solve inequalities by 'doing the same to both sides'.

These **skills boosts** will help you solve inequalities and show inequalities on a number line.

1 Solving one-sided inequalities   **2** Solving two-sided inequalities

You might have already done some work on inequalities. Before starting the first skills boost, rate your confidence using each method.

① Solve $x + 7 \geqslant 10$

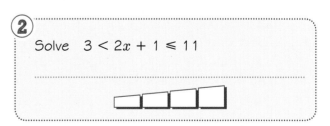

② Solve $3 < 2x + 1 \leqslant 11$

How confident are you?

1 Solving one-sided inequalities

You can solve an inequality in the same way as a linear equation, by 'doing the same to both sides'. The solution is another inequality, like $x > 3$ or $x \leq 2$. You can show this on a number line.

Guided practice

Solve $3x + 2 \leq 11$

Show the solution set on a number line.

Solve $3x + 2 \leq 11$

$3x \leq \underline{\quad}$

$x \leq 3$

Do the same to both sides to get x on its own.

Complete the number line to show possible values for x.

0 1 2 3 4 x

Use a filled circle for $\leq$ or $\geq$.
Leave it empty for $<$ or $>$.

 (1) Draw number lines to show the inequalities.

a $x \geq 3$ **b** $x < 2\frac{1}{2}$ **c** $x > -4$

x x x

 (2) Solve the inequalities and show each solution set on a number line.

a $2x + 5 \leq 9$ **b** $3x - 7 > 5$

Exam-style question

(3) Solve $\frac{x}{4} + 3 < 5$

(2 marks)

 (4) Find the integer values that satisfy

a $x > 5$ and $x \leq 8$

Hint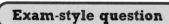
4 5 6 7 8 x

b $x \geq 2$ and $x < 7$

c $x + 3 > -2$ and $x \leq 0$

Hint Solve $x + 3 > -2$ first.

5 Solve

a $2(x + 7) < 20$

Hint Expand the bracket first.

....

b $3(n - 2) \geqslant 12$

....

6 Solve $3(y + 2) > 8$
Show the solution set on a number line.

(3 marks)

7 Divide both sides by −2. Write the correct sign in the answer.

a −4 < −2

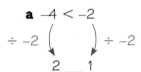

2 1

b −2 < 6

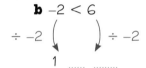

1 ....

c −2x ≤ 10

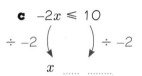

x ....

8 What happens to the sign when you divide both sides of an inequality by a negative number?

....

9 Solve

a $3x + 2 \leqslant 5x$

$2 \leqslant$ ....

.... $\leqslant x$

Hint Subtract $3x$ from both sides.

b $7x - 3 < 2x + 7$

....

10 **a** Solve $6x - 7 \geqslant 3x + 8$

.... (2 marks)

b Show the solution set on a number line.

(1 mark)

Reflect   In Q5a, could you have started by dividing both sides by 2, instead of expanding the brackets? Would this method work for Q5b and Q6?

2 Solving two-sided inequalities

You can solve a two-sided inequality by 'doing the same to all the parts'.
The solution is another two-sided inequality, like $-2 < x \leqslant 3$. You can show this on a number line.

Guided practice

Worked exam question

a Solve $-3 \leqslant 2x + 1 < 4$

Show the solution set on a number line.

b Write down the possible integer values of x.

---

a $-3 \leqslant 2x + 1 < 4$ Do the same to all parts to get x on its own.

$-4 \leqslant 2x < \text{........}$

$\text{........} \leqslant x < \dfrac{3}{2}$

Complete the number line to show possible values of x.

b $-2, -1, 0, 1$ Write down the integer values.

(1) Write down the inequalities shown on the number lines.

a
```
   o——————•
  -1  0  1  2  3  x
```
$0 \underline{\quad} x \underline{\quad} 3$

b
```
   •——————o
 -4 -3 -2 -1  0  1  2  x
```
....

c
```
   o——————•
   0  1  2  3  4  5  x
```
....

(2) Draw number lines to show the inequalities.

a $-1 < x \leqslant 3$ **b** $2 \leqslant x \leqslant 4$ **c** $-2 \leqslant x < 0$

```
  |  |  |  |  |  |  |  |
                      x
```
```
  |  |  |  |  |  |  |  |
                      x
```
```
  |  |  |  |  |  |  |  |
                      x
```

Exam-style question

(3) **a** Solve $-1 \leqslant 2x + 3 < 8$

Show the solution set on a number line.   .... **(3 marks)**

b Write down the possible integer values of x.   .... **(1 mark)**

(4) Divide all the parts by –2. Write the correct signs.

a $-6 < -4x < 2$ **b** $-8 \leqslant -2x < 4$ **c** $-12 \leqslant -2x < 6$

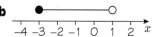

$\div -2 \left(\right) \div -2$ $\div -2 \left(\right) \div -2$

$3 \underline{\quad} 2x \underline{\quad}$ ....

$-1 \underline{\quad} 2x \underline{\quad} 3$ $\underline{\quad} < \underline{\quad} \leqslant \underline{\quad}$

Reflect   What happens to the inequality signs when you divide all the parts by a negative number?

Practise the methods

Answer this question to check where to start.

Check up

Tick the correct method for solving the inequality $-5 < 2x - 1$

A

$$-6 < 2x$$
$$-3 < x$$ ○

B

$$-4 < 2x$$
$$-2 < x$$ ○

C

$$-4 < 2x$$
$$-2 > x$$ ○

If you ticked B, go to Q3.   If you ticked A or C, go to Q1 for more practice.

1 Solve

a $7 < 4x - 1$

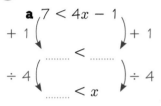

$+1 \big(\qquad\big) +1$

$\ldots < \ldots$

$\div 4 \big(\qquad\big) \div 4$

$\ldots < x$

b $-2 \leqslant 3x - 2$

$\ldots \leqslant \ldots$

$\ldots \leqslant \ldots$

c $-4 < 5x + 1$

$\ldots < \ldots$

$\ldots < \ldots$

2 Solve

a $-3x < 15$

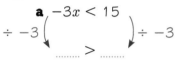

$\div -3 \big(\qquad\big) \div -3$

$\ldots > \ldots$

b $6 \leqslant -2x$

$\ldots\ldots\ldots\ldots$

c $-5x < 10$

$\ldots\ldots\ldots\ldots$

3 Solve the inequalities and show each solution set on a number line.

a $3x + 5 < 8$

b $5(x - 3) > 15$

c $1 < 3x + 2$

$\ldots\ldots\ldots$ $\ldots\ldots\ldots$ $\ldots\ldots\ldots$

4 Solve the inequalities and show each solution set on a number line.
Write down the possible integer values of x.

a $-1 \leqslant x + 1 < 5$

b $0 < -x + 3 < 7$

$\ldots\ldots\ldots$ $\ldots\ldots\ldots$

Exam-style question

5 **a** Solve $-2 \leqslant 3n < 6$ $\ldots\ldots\ldots\ldots$ (2 marks)

b Show the solution set on the number line.

$\xleftarrow{\quad}$ -2 -1 0 1 2 3 4 5 6 n (1 mark)

c Write down the possible integer values of n. $\ldots\ldots\ldots\ldots$ (1 mark)

6 Solve the inequalities. Write down the possible integer solutions.

a $-5 < 2x + 7 \leqslant 3$

b $-6 \leqslant \dfrac{x}{2} - 3 < 1$

$\ldots\ldots\ldots\ldots$ $\ldots\ldots\ldots\ldots$

Problem-solve!

Exam-style question

1 **a** Write the inequality shown in the diagram.

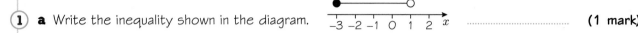
$-3\ -2\ -1\ \ 0\ \ 1\ \ 2\ \ x$

.... (1 mark)

b m is an integer.

$1 < m \leqslant 5$

List the possible integer values of m.

.... (2 marks)

c Solve $\dfrac{a}{4} + 1 > 3$

.... (2 marks)

2 Solve

a $5x > 3x + 2$

b $2x - 5 > 3x - 7$

Exam-style questions

3 Solve $5x - 3 \geqslant 2x + 7$

.... (3 marks)

4 The perimeter of the square is less than the perimeter of the triangle.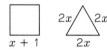

a Write an inequality to show this.

.... (1 mark)

b x is an integer.

Show that x cannot be less than 3.

.... (3 marks)

5 Mary and Bill have n counters each.
Mary loses half of her counters.
Bill loses 6 of his counters.
Bill now has more counters than Mary.
Find the smallest possible value of n.

.... (4 marks)

6 Find all the possible integer solutions to
$-2 < 3x - 5 \leqslant 3$

.... (3 marks)

7 Show that the inequality $20 < 4x^{-3} < 24$ has no integer solutions.

Now that you have completed this unit, how confident do you feel?

1 Solving one-sided inequalities

2 Solving two-sided inequalities

(11) Direct and inverse proportion

This unit will help you recognise direct and inverse proportion and write formulae for quantities in direct or inverse proportion.

AO1 Fluency check

(1) **a** $y = 3x$ **i** Find y when $x = 4$ ....   **ii** Find x when $y = 15$ ....

   b $t = \frac{1}{4}u$ **i** Find t when $u = 20$ ....   **ii** Find u when $t = 1.5$ ....

(2) Make y the subject.

   a $\frac{y}{x} = 6$ **b** $\frac{y}{x} = \frac{1}{2}$ **c** $\frac{x}{y} = 3$ **d** $\frac{x}{y} = \frac{1}{5}$

(3) **Number sense**

Circle the odd one out. $\frac{1}{3}$ $\frac{5}{15}$ $\frac{7}{24}$ $\frac{10}{30}$ $\frac{9}{27}$

Key points

↓

When x and y are in direct proportion: $y = $ a number $\times x$ or $y = kx$, where k is a number.

These **skills boosts** will help you recognise direct and inverse proportion and write formulae for quantities in direct or inverse proportion.

| (1) Recognising direct proportion | (2) Writing and using formulae for quantities in direct proportion | (3) Writing and using formulae for quantities in inverse proportion |

You might have already done some work on direct and inverse proportion. Before starting the first skills boost, rate your confidence using each method.

Use this table of values for Q1 and Q2.

x	0	2	4	6	8
y	0	4	8	12	16

(1) Show that x and y are in direct proportion.

(2) Write a formula connecting x and y.

(3) y is in inverse proportion to x. When $y = 5$, $x = 1$ Write a formula connecting y and x.

How confident are you?

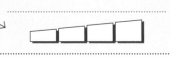 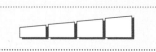 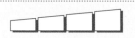

Skills boost

1 Recognising direct proportion

You can show that two quantities x and y are in direct proportion by:

showing that when one quantity is 0, the other is 0 **AND** showing that all the pairs of values $\frac{y}{x}$ (except $\frac{0}{0}$) simplify to the same value

OR showing that the graph is a straight line through (0, 0).

Guided practice

The table shows the prices for different numbers of ice creams.

a Are P and n in direct proportion? Explain your reasoning.

b Write a formula linking P and n.

Number of ice creams, n	1	2	3	4	5
Price, P (£)	2	4	6	8	10

a 0 ice creams cost £0, so when $n = 0$, $P = 0$

Show that when $n = 0$, $P = 0$ (or when $P = 0$, $n = 0$).

n	1	2	3	4	5
P	2	4	6	8	10
$\frac{P}{n}$	$\frac{2}{1}$	$\frac{4}{2}$	$\frac{6}{\ldots}$	$\frac{\ldots}{4}$	$\frac{\ldots}{\ldots}$
	$= 2$	$= \ldots$	$= \ldots$	$= \ldots$	$= \ldots$

Find $\frac{P}{n}$ for all pairs.

$\frac{P}{n}$ always simplifies to the same value.

b $\frac{P}{n} = \ldots$ so $P = 2n$

Multiply both sides by n.

① Show that all the pairs of values (except $\frac{0}{0}$) simplify to the same value.

a

x	0	1	2	3	4
y	0	3	6	9	12
$\frac{y}{x}$			...	...	...

b

e	2	4	6	8	10
f	1	2	3	4	5
$\frac{f}{e}$	...	...	...	...	...

② Write formulae linking the variables in Q1.　**a** $y = $ ....　**b** $f = $ ....

Exam-style question

③ The table shows miles converted to kilometres.

Miles (m)	0	10	100
Km (k)	0	16	160

a Write a formula connecting kilometres and miles. .... **(1 mark)**

b Convert **i** 40 miles to km .... **(1 mark)**

ii 90 km to miles (to the nearest mile). .... **(1 mark)**

Reflect Which of these formulae show quantities in direct proportion?

$y = 6x$　　$y = 2x + 4$　　$y = 3x^2$　　$y = 0.2x$

Hint Think about their graphs.

2 Writing and using formulae for quantities in direct proportion

The sign ∝ means 'is in direct proportion to'.

$y \propto x$ means 'y is in direct proportion to x' and $y = kx$, where k is a number.

These mean the same:

• y is in direct proportion to x • $y \propto x$

Guided practice

m is proportional to t. When $m = 5$, $t = 2$

a Write 'm is proportional to t' in algebra.

b Write a formula connecting m and t.

c Find m when $t = 4$

a $m \propto t$ Use the ∝ symbol.

$m = kt$ Rewrite with '=' and 'k'.

b When $m = 5$, $t = 2$

$5 = k \times 2$ Substitute the given values for m and t.

Solve for k. $\underline{} = k = \underline{}$ Write k as a decimal.

Write the value of k in the formula.

$m = \underline{} t$

c When $t = 4$

$m = 2.5 \times \underline{}$ Substitute $t = 4$ in the formula.

$m = \underline{}$

① y is proportional to x. When $y = 72$, $x = 9$ **Hint** Substitute the y and x values to find k.

  a Write 'y is proportional   **b** Write a formula.   **c** Find y when $x = 11$
    to x' in algebra.   connecting y and x.

                        $y = $

② h is directly proportional to n. When $h = 18$, $n = 10$ **Hint** Start with $h \propto n$.

  a Write a formula   **b** Find h when $n = 15$ **c** Find n when $h = 36$
    connecting h and n.

Exam-style question

③ s is directly proportional to t. When $s = 5$, $t = 1$

  a Write a formula for s in terms of t.   (1 mark)

  b Find t when $s = 0.6$ (1 mark)

Reflect   In your formulae in Q1–Q3, if you double one quantity what happens to the other quantity?

③ Writing and using formulae for quantities in inverse proportion

These mean the same:
- y is inversely proportional to x
- $y \propto \dfrac{1}{x}$

Number × Inverse = 1

$2 \times \dfrac{1}{2} = 1$

$3 \times \dfrac{1}{3} = 1$

$x \times \dfrac{1}{x} = 1$ inverse of x

Guided practice

Worked exam question

n is inversely proportional to r. When $n = 4$, $r = 2$

a Write 'n is inversely proportional to r' in algebra.

b Write a formula connecting n and r.

---

a $n \propto \dfrac{1}{r}$

$n = \dfrac{k}{r}$

b When $n = 4$, $r = 2$

$4 = \dfrac{k}{2}$ Substitute the given values for n and r.

Solve for k.

$\underline{} \times 4 = k$

$\underline{} = k$

Write the value of k in the formula.

$n = \dfrac{\underline{}}{r}$

① y is inversely proportional to t. When $y = 4$, $t = 5$

  a Write 'y is inversely proportional to t' in algebra.

  b Write a formula connecting y and t.

         $y =$ ....

  c Find y when $t = 10$

② x is inversely proportional to h. When $x = \dfrac{1}{2}$, $h = 10$

  a Write 'x is inversely proportional to h' using algebra.

  b Write a formula connecting x and h.

         $x =$ ....

  c Find x when $h = 25$

Exam-style question

③ c is in inverse proportion to m. When $m = 3$, $c = 5$

  a Write a formula for c in terms of m.   .... **(1 mark)**

  b Find m when $c = 30$ .... **(1 mark)**

Reflect   In your formulae in Q1–Q3, if you double one quantity what happens to the other quantity?

Practise the methods

Answer this question to check where to start.

Check up

'R is in direct proportion to T.' Which statement is *not* true?

 A ○

$$R \propto T$$

 B ○

$$R = \frac{k}{T}$$

 C ○

$$R = kT$$

D ○

The graph of R against T is a straight line through (0,0).

If you ticked B that is correct. Now go to Q2.

If you ticked A, C or D go to Q1 for more practice.

(1) Circle the statements for inverse proportion.

$$m \propto \frac{1}{t} \qquad m = \frac{k}{t} \qquad r = kt \qquad s \propto \frac{1}{r} \qquad t = ks \qquad p \propto n \qquad s = \frac{k}{r}$$

Hint In inverse proportion, $y \propto \frac{1}{x}$.

(2) The table shows litres (l) converted to pints (p).

Litres (l)	0	1	2	4	10
Pints (p)	0	1.75	3.5	7	17.5

a Are l and p in direct proportion? Explain your reasoning.

....

b Write a formula for l in terms of p. ....

c Convert **i** 8 litres to pints ....

          ii 28 pints to litres. ....

Exam-style questions

(3) d is directly proportional to m. When $d = 36$, $m = 3$

    a Write a formula for d in terms of m.   .... **(1 mark)**

    b Find d when $m = \frac{1}{4}$ .... **(1 mark)**

    c Find m when $d = 60$ .... **(1 mark)**

(4) x is inversely proportional to a. $x = 4$ when $a = 10$

    a Write a formula for x in terms of a.   .... **(1 mark)**

    b Find x when $a = 5$ .... **(1 mark)**

    c Find a when $x = 80$ .... **(1 mark)**

Problem-solve!

① x and y are in direct proportion.
Complete the table of values.

x	....	1	2	....	10
y	0	....	7	28	....

Exam-style questions

② The number of euros (x) is directly proportional to the number of pounds (y).
One day, 1 euro = £0.87

 a Write a formula linking x and y.

.... **(1 mark)**

 b Use your formula to convert 26 euros into pounds. .... **(1 mark)**

 c In the UK, a phone costs £89.
  In France, the same phone costs 125 euros.
  The phone is cheaper in the UK. How much cheaper?
  Give your answer in pounds or in euros.

.... **(2 marks)**

③ s is directly proportional to r. When $r = 10$, $s = 3$
Calculate r when $s = 150$

.... **(2 marks)**

④ U and V are in inverse proportion.
Complete the table of values.

U	0.5	....	....	2	3
V	....	6	4	3	....

(4 marks)

⑤ In a factory, the time (t) taken to assemble a car is inversely proportional
to the number of people (n) working on it.
5 people assemble a car in 1 hour.
How long would it take for

 a 10 people .... **(1 mark)**

 b 4 people? .... **(1 mark)**

Now that you have completed this unit, how confident do you feel?

1 Recognising direct proportion

2 Writing and using formulae for quantities in direct proportion

3 Writing and using formulae for quantities in inverse proportion

Answers

Unit 1 Mixed numbers

A01 Fluency check

① **a** $\frac{5}{8}$ **b** $\frac{1}{2}$ **c** $3\frac{13}{21}$ **d** $1\frac{19}{40}$

② **a** $\frac{2}{15}$ **b** $\frac{15}{14}$ **c** $\frac{8}{15}$ **d** $\frac{7}{12}$

③ **a** $\frac{13}{4}$ **b** $\frac{17}{6}$ **c** $\frac{22}{5}$ **d** $\frac{16}{3}$

   e $2\frac{2}{3}$ **f** $1\frac{4}{7}$ **g** $4\frac{2}{5}$ **h** $3\frac{3}{4}$

④ Number sense

   a 28 **b** 56 **c** 28

   d 7 **e** $\frac{1}{3}$ **f** 42

Confidence questions

① $4\frac{1}{4}$ ② $\frac{7}{8}$ ③ $5\frac{1}{3}$

Skills boost 1 Adding and subtracting mixed numbers

Guided practice

$= \frac{9}{4} + \frac{7}{2}$

$= \frac{9}{4} + \frac{14}{4}$

$= \frac{23}{4}$

$= 5\frac{3}{4}$

① **a** $4\frac{3}{8}$ **b** $6\frac{7}{30}$ **c** $4\frac{19}{35}$

② **a** $2\frac{3}{8}$ **b** $2\frac{1}{2}$ **c** $2\frac{1}{24}$

③ **a** $2\frac{3}{20}$ **b** $2\frac{1}{14}$

④ **a** $3\frac{3}{8}$ **b** $1\frac{3}{5}$ **c** $1\frac{4}{9}$

⑤ $12\frac{7}{8}$ kg

Skills boost 2 Multiplying mixed numbers

Guided practice

$= \frac{3}{2} \times \frac{3}{4}$

$= \frac{3 \times 3}{2 \times 4}$

$= \frac{9}{8}$

$= 1\frac{1}{8}$

① **a** $1\frac{11}{24}$ **b** $\frac{15}{16}$ **c** $1\frac{1}{5}$

② **a** $2\frac{5}{8}$ **b** $2\frac{3}{4}$ **c** $20\frac{1}{6}$

③ $4\frac{15}{32}$ square inches

④ $5\frac{5}{8}$ square metres

Skills boost 3 Dividing mixed numbers

Guided practice

$= \frac{13}{4} \div \frac{5}{2}$

$= \frac{13}{4} \times \frac{2}{5}$

$= \frac{26}{20}$

$= 1\frac{6}{20}$

$= 1\frac{3}{10}$

① **a** $2\frac{2}{15}$ **b** $3\frac{2}{3}$ **c** $2\frac{18}{25}$

② **a** $2\frac{1}{6}$ **b** $2\frac{3}{14}$ **c** $\frac{7}{10}$

③ $1\frac{1}{4}$ cm

Practise the methods

① **a** 5 **b** 8

② **a** 12 **b** 13

③ **a** $2\frac{3}{8}$ **b** $3\frac{1}{2}$ **c** $5\frac{3}{10}$

④ **a** $2\frac{2}{15}$ **b** $3\frac{2}{3}$ **c** $2\frac{18}{25}$

⑤ **a** $4\frac{1}{2}$ **b** $4\frac{1}{8}$ **c** $3\frac{1}{2}$

⑥ **a** $4\frac{13}{20}$ **b** $4\frac{3}{14}$ **c** $5\frac{1}{24}$

⑦ **a** $2\frac{1}{2}$ **b** $2\frac{1}{6}$ **c** $1\frac{3}{8}$

⑧ **a** $3\frac{19}{40}$ **b** $1\frac{17}{42}$ **c** $3\frac{13}{24}$

⑨ **a** $5\frac{1}{4}$ **b** $3\frac{1}{3}$ **c** $4\frac{2}{5}$

⑩ **a** $3\frac{1}{3}$ **b** $\frac{2}{3}$ **c** $1\frac{2}{5}$

⑪ **a** $1\frac{11}{24}$ **b** $2\frac{4}{15}$

Problem-solve!

① No. $20 \times 1\frac{7}{8} = 37\frac{1}{2}$, which is greater than 34

② $1\frac{13}{24}$ yards

③ $\frac{9}{16}$ inch

④ **a** $21\frac{1}{8}$ kg

   b 8

⑤ $55\frac{11}{16}$ or $\frac{891}{16}$ square inches

⑥ $1\frac{2}{5} \div 6\frac{4}{3} = \frac{21}{110}$

Unit 2 Brackets

AO1 Fluency check

1. **a** $2x + 6$ **b** $-4x - 4$ **c** $x^2 + 3x$
2. **a** $6x$ **b** $x^2 + 2x$ **c** $x^2 - 3x - 3$
3. **a** x **b** x **c** x
4. **a** 3 **b** y **c** x
5. **a** $4(x + 3)$ **b** $2(x + 5)$ **c** $2(x - 3)$

6 Number sense

$3 \times 5 = 15$
$3 \times -5 = -15$
$-3 \times -5 = 15$

Confidence questions

1. $x(3 + t)$
2. $4x(x + 2)$
3. $x^2 + 7x + 10$
4. $(x + 1)(x + 3)$

Skills boost 1 Factorising when the HCF is a letter or a number

Guided practice

The HCF of $2x$ and xy is x.
$2x + xy = x(\mathbf{2} + \mathbf{y})$
Check your answer by expanding.
$= \mathbf{2x} + \mathbf{xy}$

1. **a** $x(6 + t)$ **b** $x(5 - y)$ **c** $y(y + 1)$
   d $x(4 + y)$ **e** $x(5 + x)$ **f** $a(3 - b)$
2. **a** $x(xy + 1)$ **b** $cd(c + 1)$ **c** $st(t - 1)$
3. **a** $xy(x + y)$ **b** $yz(z - y)$ **c** $a^2b(b^2 + 1)$
   d $d^2(e + f)$ **e** $u^3v(v - 1)$ **f** $g^2h^2(1 + g)$
4. **a** $5(x + 4)$ **b** $x(x - 3)$ **c** $xy(x + 1)$

Skills boost 2 Factorising when the HCF has numbers and letters

Guided practice

a $3x^2 + 6x = \mathbf{3x}\,(\quad + \quad)$
$3x^2 + 6x = 3x\,(\mathbf{x} + \mathbf{2})$
Check your answer by expanding.
$= \mathbf{3x^2} + \mathbf{6x}$

b $5x^2y - 10xy = \mathbf{5xy}(\quad - \quad)$
$5x^2y - 10xy = 5xy(\mathbf{x} - \mathbf{2})$
Check your answer by expanding.
$= \mathbf{5x^2y} - \mathbf{10xy}$

1. **a** $4x(x + 2)$ **b** $3b(c - 4b)$
   c $2x(3x + 2y)$ **d** $3x(2x - 1)$
   e $5m(m - 4)$ **f** $2d(a + 5d)$
2. **a** $7p(q - 2p + 2q^2)$
   b $3y(1 + 5x^2y - 3xz)$
3. **a** $4x^2$ and $6xy$ have common factor $2x$
   b $2x$ and $6y$ have common factor 2
   c x^2 and $3xy$ have common factor x
4. **a** $4x(2x - 3y)$ **b** $5ac(b + 3a)$
   c $9xy(1 + 3y)$ **d** $6ef(e - 3g)$
   e $10xy(2x - 3y)$ **f** $6b^2(a - 5)$
5. **a** $8wz(wz - 3v^2 + 2vw)$
   b $3b^2(2ac - ac^2 - 4c)$
6. **a** $8(x - 2)$
   b $6xy^2(1 + 2x)$

Skills boost 3 Expanding double brackets

Guided practice

$= x^2 + 5x + \mathbf{3x} + \mathbf{15}$
$= x^2 + \mathbf{8x} + 15$

1. **a** $x^2 + 6x + 8$ **b** $x^2 + 2x - 3$
   c $x^2 + 3x - 10$ **d** $x^2 - x - 6$
   e $m^2 - 9m + 14$ **f** $y^2 - 5y + 4$
2. $m^2 + 13m + 40$
3. **a** $x^2 - 1$ **b** $x^2 - 4$ **c** $b^2 - 25$
4. **a** $x^2 - 9$ **b** $a^2 - 100$
5. **a** $x^2 + 6x + 9$ **b** $x^2 + 8x + 16$
   c $x^2 + 10x + 25$
6. Students' own workings
7. Students' own workings

Skills boost 4 Factorising quadratic expressions

1. **a** $(x + 1)(x + 3)$ **b** $(x - 1)(x + 7)$
   c $(x + 1)(x - 6)$
2. **a** $(x - 1)(x + 5)$ **b** $(x - 2)(x + 3)$
   c $(x - 1)(x - 3)$
3. **a** $(x - 5)(x + 5)$ **b** $(x - 9)(x + 9)$
4. **a** $(x - 10)(x + 10)$ **b** $(x - 4)(x - 4)$

Practise the methods

1. **a** $2x(x - 2y)$ **b** $3a(2b + 3a)$
2. **a** $4x^2$ and $6xy$ have common factor $2x$
   b $4x$ and $6y$ have common factor 2
   c $2x^2$ and $3xy$ have common factor x
3. **a** $x(3 + z)$ **b** $a(a - 1)$
   c $b(2 + b)$ **d** $mp(m + 1)$
   e $2x(y + 3x)$ **f** $3c(d^2 - 4c)$
4. **a** $10x(x + 2y)$ **b** $4c(4b - cd)$
   c $6xy(2y^2 - x)$
5. **a** $x^2 + 12x + 35$ **b** $n^2 - 5n - 14$
   c $t^2 - 11t + 24$
6. **a** $x^2 + 4x + 4$ **b** $x^2 - 81$
   c $x^2 - 16x + 64$
7. **a** $(x + 2)(x + 5)$ **b** $(x + 2)(x - 3)$
   c $(x - 1)(x - 7)$
8. **a** $(x + 4)(x - 4)$ **b** $(x - 5)^2$

Problem-solve!

1. Students' own workings
2. Students' own workings
3. $\frac{1}{2}(x + 3)(x - 1)$
   $= \frac{1}{2}(x^2 + 2x - 3)$
   $= \frac{1}{2}x^2 + x - \frac{3}{2}$
4. Students' own workings
5. $(x + 6)(x + 7) - x(x + 1) = x^2 + 13x + 42 - x^2 - x$
   $= 12x + 42$
   $= 6(2x + 7)$
6. $x(x + 5) = x^2 + 5x$
7. $\sqrt{2x^2 + 4x + 4}$

Unit 3 Formulae

AO1 Fluency check

1. **a** 7 **b** 10 **c** $-\frac{1}{2}$
   d 4 **e** 4 **f** -25

② **a** 4   **b** 5
③ **a** $x = 5$ **b** $x = 6$
④ **a** $3x$ **b** nx **c** $\dfrac{x}{2}$ **d** $2x + 10$

⑤ Number sense

$7 \times \underline{8} = 56$ $9 \times \underline{5} = 45$
$56 \div \underline{7} = 8$ $45 \div \underline{9} = 5$
$56 \div \underline{8} = 7$ $45 \div \underline{5} = 9$

Confidence questions

① $t = 2$
② $P = 8h$
③ $d = st$
④ $A = \dfrac{F}{P}$

Skills boost 1 Finding a value that is not the subject of a formula

Guided practice

$v = u + at$
$20 = u + \underline{4} \times \underline{3}$
$20 = u + \underline{12}$
$20 - \underline{12} = u$
$u = 8$

① **a** $R = 3$ **b** $I = 2$
② **a** $m = 40$ **b** $V = 3$ **c** $V = 0.5$
③ **a** $b = 4$ **b** $b = 8$ **c** $h = 2.5$
④ $3\frac{1}{2}$ hours
⑤ **a** $t = 2$ **b** $a = 2$
⑥ **a** $a = 2$ **b** $u = 5$

Skills boost 2 Deriving a simple formula

Guided practice

1 ticket: Total cost $= 5.6 \times 1$
2 tickets: Total cost $= \underline{5.6} \times 2$
n tickets: Total cost $= \underline{5.6} \times \underline{n}$
$T = 5.6n$

① $A = 9.8m$
② $T = 9.8m + 5.6n$
③ $B = 15 + 6.50h$
④ $T = 10 + 12h$
⑤ **a** $3x + 4x$ **b** $x^2 + x + 2$ **c** $2xy - x^2$
⑥ $y^3 + 6y^3$

Skills boost 3 Changing the subject of a formula

Guided practice

$2 \times A = \dfrac{{}^1\!2 \times bh}{\cancel{2}}$
$2A = bh$
Divide both sides by b.
$\dfrac{2A}{b} = \dfrac{{}^1\!bh}{\cancel{b}}$
$\dfrac{2A}{b} = h$

① **a** $I = \dfrac{V}{R}$
   b $m = VD$
② **a** $d = st$
   b $F = PA$

③ **a** $b = \dfrac{2A}{h}$
   b $u = v - at$
④ **a** $a = \dfrac{v - u}{t}$
   b $t = \dfrac{v - u}{a}$
⑤ $x = \dfrac{y - c}{m}$

Skills boost 4 Changing the subject of a more complex formula

Guided practice

$\dfrac{s \times 1}{d} = \dfrac{{}^1\!s \times t}{\cancel{s}}$
$\dfrac{s}{d} = t$

① **a** $v = \dfrac{m}{D}$ **b** $A = \dfrac{F}{P}$
② **a** $I = \dfrac{V}{R}$ **b** $d = st$
③ **a** $h = \dfrac{2A}{b}$ **b** $b = \dfrac{2A}{h}$
④ **a** $F = AP$ **b** $A = \dfrac{F}{P}$
⑤ $t = \dfrac{p}{q - x}$
⑥ **a** $x = \dfrac{b}{m - c}$ **b** $x = \dfrac{r}{y + f}$
⑦ **a** $x = \sqrt{25 + a}$ **b** $x = \sqrt{d - 16}$
   c $x = \sqrt{v + u}$ **d** $x = \sqrt{\dfrac{m}{n}}$
⑧ **a** $t = \sqrt{3m + r}$ **b** $t = \sqrt{x - k}$
   c $t = \sqrt{y + \dfrac{x}{3}}$ **d** $t = \sqrt{sn}$
⑨ $u = \sqrt{v^2 - 2as}$
⑩ **a** $y = x^2$ **b** $y = 16t^2$
   c $y = \dfrac{p^2}{n^2}$ **d** $y = 5q^2$

Practise the methods

① **a** $2(x + 3) \longleftarrow \boxed{\times 2} \longleftarrow x + 3 \longleftarrow \boxed{+ 3} \longleftarrow x$
   b $3(m + k) \longleftarrow \boxed{\times 3} \longleftarrow m + k \longleftarrow \boxed{+ k} \longleftarrow m$
   c $a(y + t) \longleftarrow \boxed{\times a} \longleftarrow y + t \longleftarrow \boxed{+ y} \longleftarrow t$
② $z = t(f + m)$
③ $x = \sqrt{\dfrac{r}{q}}$
④ **a** $s = 3$ **b** $t = 5.5$
⑤ $x = 8$
⑥ **a** $u = 2$ **b** $a = 8.5$
⑦ $C = 85 + 12.5n$

Problem-solve!

① 202.5 km/h.
② 2.448 kg
③ 2 hours 32 minutes
④ **a** 27.8 m/s²
   b 9.1 seconds
⑤ 66.7 metres (to 1 d.p.)
⑥ 6 people
⑦ $3 \times 120\,\text{cm} + 1 \times 68\,\text{cm} = £105.25$

Unit 4 Equations

AO1 Fluency check

① **a** $x = \dfrac{3}{2}$ **b** $x = 20$

   c $x = 3$

② **a** $2x + 10$ **b** $-3x - 12$

   c $4x - 1$

③ **a** $(x + 1)^2$ **b** $(x + 3)(x - 2)$

   c $(x + 3)(x - 3)$

④ Number sense

$a = 0, b = 3; a = 7, b = 0; a = -8, b = 0;$
$a = 0, b = 0; a = 0, b = -5$

Confidence questions

① $x = -2$

② $x = 12$

③ $x = 1$ or $x = -2$

④ $x = 30°$

Skills boost 1 Solving linear equations with brackets

Guided practice

$3(x + 2) = 10$
$3x + 6 = 10$
$3x = 4$
$x = \dfrac{4}{3}$

① **a** $x = \dfrac{3}{2}$ **b** $x = 3$ **c** $x = -1$

② **a** $x = -7$ **b** $x = -4$ **c** $x = \dfrac{19}{3}$

③ **a** $x = -3$ **b** $x = -\dfrac{5}{6}$ **c** $x = \dfrac{9}{11}$

④ $x = 0$

Skills boost 2 Solving linear equations with fractions

Guided practice

$\dfrac{x}{3} - 5 = 2$

$\times 3 \Big(\dfrac{x}{3} = 7 \atop x = 21 \Big) \times 3$

① **a** $x = 98$ **b** $a = -12$ **c** $b = 22$

② **a** $w = 4$ **b** $x = -3.5$ **c** $x = 1.25$

③ **a** $4x + 1$ **b** $3x - 5$ **c** $5d + 1$

④ **a** $x = \dfrac{5}{2}$ **b** $x = \dfrac{4}{9}$ **c** $t = -\dfrac{2}{7}$

⑤ $x = \dfrac{21}{26}$

Skills boost 3 Solving quadratic equations

Guided practice

$x^2 + 2x - 8 = 0$
$(x + 4)(x - 2) = 0$
So $x + 4 = 0$ or $x - 2 = 0$
$x = -4$ or $x = 2$

① **a** $x = -2$ or $x = 3$

   b $x = 3$ or $x = 6$

   c $x = -2$ or $x = -5$

② **a** $x = 2$ or $x = -2$

   b $x = -3$

   c $x = 4$ or $x = -4$

③ $m = 1$ or $m = -4$

④ **a** $x = 0$ or $x = 5$

   b $x = 0$ or $x = -7$

   c $x = 0$ or $x = \dfrac{3}{2}$

⑤ **a** $x = -5$ or $x = 4$

   b $x = -2$ or $x = -4$

   c $x = 2$ or $x = 5$

Skills boost 4 Writing equations to solve problems

Guided practice

$2x + x + 90 = 180$
$3x + 90 = 180$
$3x = 90$
$x = 60$

Angle $x = 30°$; angle $2x = 60°$

① **a** $x = 20°, 3x = 60°$

   b $x = 45°, 2x + 10 = 100°$

② $x = 3\,\text{cm}, x + 2 = 5\,\text{cm}$

③ $x = 5\,\text{cm}$

④ **a** $x = 48°, 2x - 10 = 86°, 2x + 20 = 116°$

   b $x = 45°, 3x = 135°$

⑤ $30°$

⑥ $8.5\,\text{cm}$

⑦ $2x + 30 = 110°, 3x - 10 = 110°, y = 70°,$
   unknown $= 70°$

⑧ $7\,\text{m}$

⑨ 90p

⑩ £1.70

⑪ 15, 16, 17

Practise the methods

① **a** $x = 20$ **b** $x = 6$ **c** $x = 12$

② **a** $x = \dfrac{4}{5}$ **b** $x = 18$ **c** $x = 2$

③ **a** $x = 6$ **b** $x = -\dfrac{8}{3}$ **c** $x = -\dfrac{31}{6}$

④ $x = -\dfrac{9}{10}$

⑤ **a** $x = 6, x = -6$ **b** $x = 0, x = 8$

   c $x = 0, x = -\dfrac{5}{2}$

⑥ $x = -7$ or $x = 2$

Problem-solve!

① $x = 80°, 2x = 160°$

② $7.5\,\text{cm}$

③ 6

④ £16

⑤ $x = 2$

⑥ 7

⑦ $x^2 + 2x - 15 = 0$

Unit 5 Simultaneous equations

AO1 Fluency check

① **a** x **b** $3y$ **c** $2x$ **d** $-4y$

② **a** $x = 3$ **b** $y = -4$ **c** $x = 7$ **d** $y = -\dfrac{7}{2}$

③ $x = 4$

④ Number sense

$2x + 2y = 20$

$5x + 5y = 50$

$3x + 3y = 30$

Confidence questions

① $x = 1, y = 3$

② $x = 4, y = 4$

③ $x = 2, y = 1$

④ Shirt £15, hat £8

Skills boost 1 Subtracting to eliminate a variable

> **Guided practice**

Subtract B from A.

$3y - y = \underline{2}$

$\quad y = \underline{1}$

Substitute $y = \underline{1}$ into A.

$2x + \underline{3} = 9$

$\quad 2x = \underline{6}$

$\quad\; x = 3$

① **a** $x = 1, y = 3$

    b $x = 4, y = 2$

    c $x = 2, y = 1$

② **a** $x = 3, y = 5$

    b $x = 3, y = 1$

    c $x = -2, y = 3$

③ $x = 5, y = 2$

④ **a** $x = 4, y = 1$

    b $x = 5, y = 2$

    c $x = -3, y = 19$

Skills boost 2 Adding to eliminate a variable

> **Guided practice**

Add A and B and solve to find x.

$3x + 0 = \underline{21}$

$\quad 3x = \underline{21}$

$\quad\; x = \underline{7}$

Substitute $x = 7$ into B.

$\underline{7} + y = 10$

$\quad\;\; y = \underline{3}$

① **a** $x = 3, y = 4$

    b $x = 5, y = 1$

    c $x = 10, y = 2$

② $x = 5, y = 2$

③ $x = 6, y = 3$

④ **a** $x = 3, y = 7$

    b $x = 4, y = 7$

    c $x = 5, y = 2$

Skills boost 3 Multiplying an equation first

> **Guided practice**

Multiply A by 4.

$\underline{8}x + \underline{4}y = -4$

Subtract B from 4 × A and solve to find x.

$8x + 4y = -4$

$\;x + 4y = 10$

$\underline{7}x + 0 = -14$

$\quad\;\; x = -2$

Substitute $x = -2$ into A.

$(2 \times -2) + y = -1$

$\quad\; \underline{-4} + y = -1$

$\qquad\quad y = \underline{3}$

① $x = 3, y = 4$

② **a** $x = 4, y = 2$ **b** $x = -3, y = 1$

    c $x = -5, y = -2$

③ **a** $x = -4, y = 2$ **b** $x = 7, y = -2$

    c $x = 6, y = 3$

④ **a** $x = -1, y = 9$ **b** $x = 8, y = -3$

    c $x = 10, y = 3$

⑤ $x = -1, y = 3$

⑥ **a** $x = -2, y = 1$ **b** $x = -6, y = 2$

    c $x = -2, y = 11$

⑦ **a** $x = 2, y = \frac{1}{2}$ **b** $x = \frac{2}{5}, y = -3$

    c $x = 2, y = -\frac{3}{5}$

⑧ **a** $x = \frac{3}{4}, y = -2$ **b** $x = 3, y = 4$

    c $x = \frac{1}{2}, y = 5$

⑨ $x = \frac{9}{4}, y = -2$

Skills boost 4 Setting up simultaneous equations

> **Guided practice**

Multiply B by 3.

$15x + \underline{3}y = \underline{21}$

Subtract A from 3 × B.

$15x + 3y = 21$

$\;2x + 3y = 8$

$13x + 0 = \underline{13}$

$\quad\;\; x = \underline{1}$

a A cookie costs £$\underline{1}$ **b** A sandwich costs £$\underline{2}$

① **a** £1   **b** £2

② **a** 50p   **b** £1

③ **a** £10   **b** £6

④ **a** £8   **b** £5

⑤ **a** £1.25   **b** £4

Practise the methods

① **a** $x = 10, y = 12$ **b** $x = 5, y = 2$

    c $x = 1, y = 8$

② **a** $x = 3, y = 4$ **b** $x = 7, y = \frac{3}{2}$

    c $x = \frac{1}{3}, y = -\frac{8}{3}$

③ **a** $x = 6, y = -3$ **b** $x = -4, y = 2$

    c $x = \frac{1}{2}, y = -3$

④ **a** $x = -2, y = 3$ **b** $x = -\frac{1}{2}, y = \frac{3}{4}$

    c $x = -\frac{4}{3}, y = 1\frac{2}{5}$

⑤ **a** £2.75   **b** £2.75

⑥ $x = 3, y = -5$

Problem-solve!

① **a** £9   **b** £5

② $x = 11, y = 5$

③ 10 and 25

④ $x = \frac{1}{2}, y = -5$

⑤ $h = £10, b = £30$

6 a A tea costs £1.50 and a coffee costs £2.50

b A milkshake costs £1.75 and a cookie costs £1.20

7 9p

8 It is not possible to solve these equations because they are equivalent.

Unit 6 Indices

AO1 Fluency check

1 a 25   **b** 8   **c** 16   **d** -1

2 a $-\dfrac{1}{2}$ **b** $-\dfrac{5}{7}$ **c** 4   **d** -6

3 a $\dfrac{8}{15}$ **b** $\dfrac{1}{8}$ **c** 5   **d** $\dfrac{7}{2}$

4 Number sense

a 2^2 **b** $\dfrac{1}{5}$ **c** 4^2

Confidence questions

1 3^5

2 4

3 $\dfrac{1}{8}$

Skills boost 1 Laws of indices for multiplication and division

Guided practice

a $5^3 \times 5^4 = 5^{3+4}$
$= 5^7$

b $\dfrac{4^4}{4^3} = 4^{4-3} = 4$

1 a 3^7 **b** 4^6 **c** 2^9

2 a 2^4 **b** 10^4 **c** 5^2

3 a 32   **b** 9   **c** 1000

4 a 5^4 **b** 7^2 **c** 4^3

5 4^6

Skills boost 2 Reciprocals

Guided practice

a $1 \div 5 = \dfrac{1}{5}$

b $1 \div \dfrac{1}{3}$
$= 1 \times \dfrac{3}{1}$
$= \dfrac{3}{1}$
$= 3$

1 a $\dfrac{1}{6}$ **b** 8   **c** -2

2 a $\dfrac{5}{3}$ **b** $-\dfrac{7}{5}$ **c** $\dfrac{2}{3}$

3 $3 \to \dfrac{1}{3} \to 3, \dfrac{1}{7} \to 7 \to \dfrac{1}{7}, \dfrac{3}{8} \to \dfrac{8}{3} \to \dfrac{3}{8}$

4 a $\dfrac{1}{12}$ **b** 1

5 a $\dfrac{9}{4}$ **b** 1

6 5

Skills boost 3 Zero and negative indices

Guided practice

a $8^0 = 1$

b $3^{-1} = 1 \div 3$
$= \dfrac{1}{3}$

c $\left(\dfrac{5}{7}\right)^{-1} = 1 \div \dfrac{5}{7}$
$= 1 \times \dfrac{7}{5}$
$= \dfrac{7}{5}$

1 a 1   **b** $\dfrac{1}{6}$ **c** $\dfrac{4}{3}$ **d** $\dfrac{2}{5}$

2 a 2^3 **b** 4^3 **c** 5^2 **d** 3^{-3}

3 a $\dfrac{1}{9}$ **b** $\dfrac{1}{25}$ **c** 4

4 a $\dfrac{1}{4}$ **b** $\dfrac{1}{1000}$ **c** 27

Practise the methods

1 a 5^3 **b** 7^6 **c** 4^2

2 a 3^{-1} **b** 10^{-2} **c** 5^0

3 a 3^{10} **b** 2^{18} **c** 7^{12}

4 a 64   **b** $\dfrac{5}{3}$ **c** 25

5 a -1 **b** 6   **c** 1

6 a 1   **b** $\dfrac{1}{64}$ **c** 2

   d $\dfrac{1}{16}$ **e** $\dfrac{1}{9}$ **f** 8

Problem-solve!

1 $\dfrac{5}{9} \times \dfrac{9}{5} = 1$

2 a x^7 **b** y^5 **c** z^2

3 $\dfrac{1}{2x} = (2x)^{-1}$, $\dfrac{1}{x^2} = x^{-2}$, $\dfrac{2}{x} = \left(\dfrac{x}{2}\right)^{-1}$, $x^0 = 1$

4 a a **b** m^{-4} or $\dfrac{1}{m^4}$ **c** n^3

5 x^3

6 a 0.1219512195   **b** 0.122

Unit 7 Pythagoras' theorem

AO1 Fluency check

1 a 36   **b** 89

2 a 26.96   **b** 4.277

3 65

4 a $c = 7$ **b** $x = 24$

5 A, B, D

6 Number sense

Odd one out is 8.08..., which rounds to 8.1

Confidence questions

1 13 cm

2 8.9 cm

3 3.24 m

Skills boost 1 Using Pythagoras to find the longest side in a right-angled triangle

Guided practice

$x^2 = 3^2 + 4^2$

$\quad = 9 + 16$

$\quad = 25$

$x = \sqrt{5}$

$\quad = 5\,cm$

① **a** 10 cm

　 b 8.6 cm

② **a** 5.8 cm

　 b 10.8 cm

③ 5.5 cm

④ **a** 9.2 cm

　 b 5.4 cm

Skills boost 2 Using Pythagoras to find one of the shorter sides in a right-angled triangle

Guided practice

$13^2 = 5^2 + y^2$

$169 = 25 + y^2$

Solve to find y.

$y^2 = 169 - 25$

$y = \sqrt{144}$

$y = 12\,cm$

① **a** 5.3 cm　　**b** 6.0 cm　　**c** 7.5 cm

② **a** 12.7 m　　**b** 4.8 km　　**c** 15.0 mm

③ 18.7 cm

Skills boost 3 Solving problems involving right-angled triangles

Guided practice

$3.8^2 = 2.5^2 + b^2$

$b^2 = 14.44 - 6.25$

$b = \sqrt{8.19}$

wall = 2.86 m = 286 cm

① **a**

　 b 6.4 km (to 1 d.p.)

② 12.2 km (to 1 d.p.)

③ **a** $b = 2$　　**b** 6.3 (to 1 d.p.)

④ 8.2 (to 1 d.p.)

Practise the methods

① **a** 7.5　　**b** 7.1　　**c** 8.9

② **a** 10.3 cm　　**b** 5.8 cm　　**c** 11.9 m

③ **a** 10.4 cm　　**b** 54.9 mm　　**c** 4.9 km

④ 9.2 m (to 1 d.p.)

⑤ 5.6 (to 1 d.p.)

Problem-solve!

① Yes. (It reaches 3.82 m.)

② 7.7 cm

③ 8.4 cm (to 1 d.p.)

④ No, journey is 154 km.

⑤ No, $3.6^2 + 4.2^2 \neq 5.8^2$

⑥ **a** AB (AB = 4.47 cm, AC = 4.12 cm)

　 b 0.349 cm

⑦ AB = BC = 5.7 cm (to 1 d.p.)

Unit 8 The equation of a straight line

AO1 Fluency check

① **a** 5　　　　**b** 11　　　　**c** 1

② **a** $y = 5$　　**b** $x = 2$

③ **a** $y = 4 - 3x$

　 b $y = \dfrac{-7}{2}x + \dfrac{11}{2}$

　 c $y = \dfrac{x - 9}{5}$

④ **Number sense**

　 a -5　　**b** -3　　**c** -1

Confidence questions

① $y = 2x - 2$

② Gradient 4, y-intercept -5

③ Gradient $\dfrac{-4}{3}$, y-intercept 2

Skills boost 1 Finding the equation of a straight line from its graph

Guided practice

Find the y-intercept.

$c = 2$

Find the gradient.

$m = 3$

Substitute m and c into $y = mx + c$

$y = 3x + 2$

① **a** $y = x + 4$

　 b $y = 4x + 1$

　 c $y = \dfrac{1}{2}x + 3$

② **a** $y = -x + 3$

　 b $y = -2x + 2$

　 c $y = -\dfrac{1}{3}x + 4$

③ $y = \dfrac{1}{2}x - 1$

Skills boost 2 Using the equation $y = mx + c$

Guided practice

x	-2	-1	0	1	2
y	-5	-3	-1	1	3

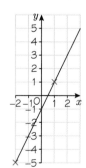

①

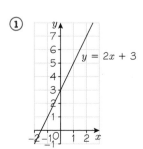

$y = 2x + 3$

② A $y = 3x$, E $y = 3x - 2$, F $y = 3x + 5$

③ **a** Students' own answers: $y = 3x + c$ for any number c

  b Students' own answers: $y = -5x + c$ for any number c

④ **a** $(0, 2)$ **b** $(0, -1)$ **c** $(0, 0)$

⑤ $y = 4x + 5$

Skills boost 3 Using the equation $ax + by = c$

Guided practice

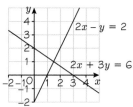

x	0	4
y	2	0

①, ②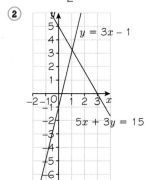

2x − y = 2
2x + 3y = 6

③ **a** $y = 2x - 2$

  b Gradient 2, y-intercept -2

④ A $-2x + y = 3$, E $2x - y = 5$, F $4x - 2y = 7$

⑤ y-intercept 6, gradient $\dfrac{-3}{2}$

Practise the methods

① **a** $y = -x - 2$

  b $y = 3x + 1$

  c $y = \dfrac{1}{2}x - 1$

②

$y = 3x - 1$
$5x + 3y = 15$

③ A and B

④ **a** Students' own answers: $y = 4x + c$ for any number c

  b Students' own answers: $3x - 6y = c$ for any number c

Problem-solve!

① A, C, D, B

② **a** A and C   **b** A and B

③ $y = 3x - 2$

④

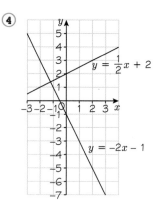

$y = \dfrac{1}{2}x + 2$

$y = -2x - 1$

⑤ $y = -2x - 1$

⑥ $x = 9, y = 2$

Unit 9 Non-linear graphs

A01 Fluency check

① **a** 9   **b** 4   **c** 8   **d** -1

② **a** 4   **b** 13   **c** 5

③ **a** -2 **b** 5   **c** -4

④ Number sense

  a 10   **b** 0   **c** 4

Confidence questions

① 5, 2, 1, 2, 5

② $x = 1, x = 3$

③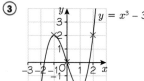

$y = x^3 - 3x$

Skills boost 1 Drawing quadratic graphs

Guided practice

x	-2	-1	0	1	2	3
y	3	0	-1	0	3	8

$y = x^2 - 1$

① **a**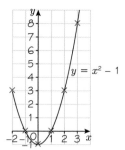

$y = x^2$

b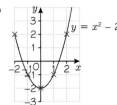

②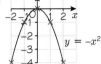

Skills boost 2 Finding solutions and turning points from quadratic graphs

Guided practice

a $y = -2\frac{1}{4}$

b $x = \underline{-1}$ and $x = \underline{2}$

① **a** $y = -4$ **b** $x = -2$ and $x = 2$

② **a** $y = 0$ **b** $x = 1$

③ **a** $\left(\frac{1}{2}, -4\right)$ **b** $x = 1\frac{1}{2}$ and $x = -\frac{1}{2}$

Skills boost 3 Drawing cubic graphs

Guided practice

a

x	-2	-1	0	1	2
y	$\underline{-9}$	-2	-1	0	$\underline{7}$

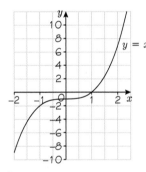

b $y = 2.4$

① **a**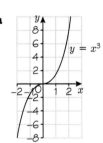

b $y \approx 2.7$

② **a**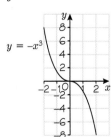

b ≈ -1.7

① $-3, 0, 1, 0, -3$

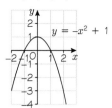

② **a** $y = -4$

    b $x = -1$ and $x = 3$

③ $-26, -7, 0, 1, 2, 9, 28$

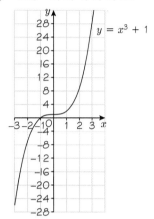

④ A ii   B iii   C iv   D i

Problem-solve!

① **a**

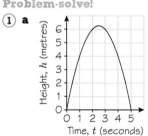

    b $\approx 6.25\,\text{m}$

② **a**

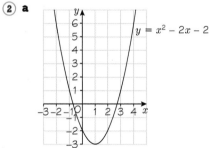

    b $(1, -3)$ **c** $x = 1$

③ **a** $y \approx 2.5$ **b** $y \approx 1.7$ **c** $x \approx -1.25$

Unit 10 Inequalities

A01 Fluency check

① **a** $-2, -1, 0, 1.8$

    b $-2, -1, 0$

② **a** $x = 10$ **b** $x = -2$ **c** $x = 1$

③ **a** $x > 1$ **b** $x \leqslant -2$ **c** $x \leqslant 5$

④ Number sense

  a $4 < 7$
  b $-4 > -7$
  c $-4 < 7$
  d $4 > -7$

Confidence questions

① $x \geqslant 3$
② $1 < x \leqslant 5$

Skills boost 1 Solving one-sided inequalities

Guided practice

$3x + 2 \leqslant 11$
   $3x \leqslant \underline{9}$
    $x \leqslant 3$

① **a**

  b $x < 2\frac{1}{2}$

  c $x > -4$

② **a** $x \leqslant 2$

  b $x > 4$

③ $x < 8$
④ **a** $6, 7, 8$
  b $2, 3, 4, 5, 6$
  c $-4, -3, -2, -1, 0$
⑤ **a** $x < 3$
  b $n \geqslant 6$
⑥ $y > \dfrac{2}{3}$

⑦ **a** $2 > 1$ **b** $1 > -3$ **c** $x \geqslant 5$
⑧ When you divide both sides of an inequality by a negative number you reverse the inequality sign.
⑨ **a** $2 \leqslant 2x$
   $1 \leqslant x$
  b $x < 2$
⑩ **a** $x \geqslant 5$
  b

Skills boost 2 Solving two-sided inequalities

Guided practice

a $-3 \leqslant 2x + 1 < 4$
  $-4 \leqslant 2x < \underline{3}$

  $\underline{-2} \leqslant x < \dfrac{3}{2}$

b $-2, -1, 0, 1$
① **a** $0 < x \leqslant 3$ **b** $-3 \leqslant x < 1$ **c** $1 < x \leqslant 3\frac{1}{2}$
② **a** $-1 < x \leqslant 3$

  b $2 \leqslant x \leqslant 4$

  c $-2 \leqslant x < 0$

③ **a** $-2 \leqslant x < \dfrac{5}{2}$

  b $-2, -1, 0, 1, 2$
④ **a** $3 > 2x > -1$
   $-1 < 2x < 3$
  b $4 \geqslant x > -2$
   $-2 < x \leqslant 4$
  c $6 \geqslant x > -3$

Practise the methods

① **a** $\underline{8} < \underline{4}x$
   $\underline{2} < x$
  b $0 \leqslant 3x$
   $0 \leqslant x$
  c $-5 < 5x$
   $-1 < x$
② **a** $x > -5$
  b $-3 \geqslant x$
  c $x > -2$
③ **a** $x < 1$

  b $x > 6$

  c $x > -\dfrac{1}{3}$

④ **a** $-2 \leqslant x < 4$

  $-2, -1, 0, 1, 2, 3$

b $-3 < x < 4$

$-3 < x < 4$

$-2, -1, 0, 1, 2, 3$

5 a $-\frac{2}{3} \leq n < 2$

b $-\frac{2}{3} \leq n < 2$

c $0, 1$

6 a $-6 < x \leq -2$, $-5, -4, -3, -2$

b $-2 \leq x < 8$, $-2, -1, 0, 1, 2, 3, 4, 5, 6, 7$

Problem-solve!

1 a $-3 \leq x < 1$

b $2, 3, 4, 5$

c $a > 8$

2 a $x > 1$

b $x < 2$

3 $x \geq \frac{10}{3}$

4 a $4(x + 1) < 6x$

b $x > 2$, and x is an integer so $x \geq 3$

5 $\frac{n}{2} < n - 6$, so $n > 12$. n cannot be 13 as Mary cannot lose $13 \div 2 = 6\frac{1}{2}$ counters. So $n = 14$.

6 $1 < x \leq \frac{8}{3}$, so $x = 2$ only.

7 $5 < x^{-3} < 6$

There are no integer solutions as x^{-3} is only greater than 5 or less than 6.

Unit 11 Direct and inverse proportion

AO1 Fluency check

1 a i 12 **ii** 5

b i 5 **ii** 6

2 a $y = 6x$ **b** $y = \frac{1}{2}x$

c $y = \frac{1}{3}x$ **d** $y = 5x$

3 Number sense

$\frac{7}{24}$

Confidence questions

1 The pairs of values of x and y are all in the same ratio. When $x = 0$, $y = 0$, and for other values $\frac{y}{x}$ is always 2

2 $y = 2x$

3 $y = \frac{5}{x}$

Skills boost 1 Recognising direct proportion

Guided practice

a

n	1	2	3	4	5
P	2	4	6	8	10
$\frac{P}{n}$	$\frac{2}{1}$ $= 2$	$\frac{4}{2}$ $= 2$	$\frac{6}{3}$ $= 2$	$\frac{8}{4}$ $= 2$	$\frac{10}{5}$ $= 2$

b $\frac{P}{n} = 2$ so $P = 2n$

1 a

$\frac{y}{x}$		$\frac{3}{1} = 3$	$\frac{6}{2} = 3$	$\frac{9}{3} = 3$	$\frac{12}{4} = 3$

b

$\frac{f}{e}$	$\frac{1}{2}$	$\frac{2}{4} = \frac{1}{2}$	$\frac{3}{6} = \frac{1}{2}$	$\frac{4}{8} = \frac{1}{2}$	$\frac{5}{10} = \frac{1}{2}$

2 a $y = 3x$

b $f = \frac{1}{2}e$

3 a $k = \frac{16}{10}m$ or $\frac{8}{5}m$ or $1.6m$

b i 64 km **ii** 56 miles

Skills boost 2 Writing and using formulae for quantities in direct proportion

Guided practice

a $m = kt$

b $5 = k \times 2$

$\frac{5}{2} = k = 2.5$

$m = 2.5t$

c $m = 2.5 \times 4$

$m = 10$

1 a $y \propto x$, $y = kx$ **b** $y = 8x$ **c** $y = 88$

2 a $h = 1.8n$ **b** $h = 27$ **c** $n = 20$

3 a $s = 0.2t$ **b** $t = 3$

Skills boost 3 Writing and using formulae for quantities in inverse proportion

Guided practice

a $n = \frac{k}{r}$

b $4 = \frac{k}{2}$

$2 \times 4 = k$

$8 = k$

$n = \frac{8}{r}$

1 a $y \propto \frac{1}{t}$, $y = \frac{k}{t}$ **b** $y = \frac{20}{t}$ **c** $y = 2$

2 a $x \propto \frac{1}{h}$, $x = \frac{k}{h}$ **b** $x = \frac{5}{h}$ **c** $x = \frac{1}{5}$

3 a $c = \frac{15}{m}$ **b** $m = \frac{1}{2}$

Practise the methods

1 $m \propto \frac{1}{t}$ $m = \frac{k}{t}$ $s \propto \frac{1}{r}$ $s = \frac{k}{r}$

2 a Yes, either graph sketched/drawn through (0, 0) or all ratios $\frac{1}{1.75}$, $\frac{2}{3.5}$ etc simplify to $\frac{1}{1.75}$

b $l = 1.75p$

c i 14 pints **ii** 16 litres

③ **a** $d = 12m$ **b** $d = 3$ **c** $m = 5$

④ **a** $x = \dfrac{40}{a}$ **b** $x = 8$ **c** $a = \dfrac{1}{2}$

Problem-solve!

①

x	0	1	2	8	10
y	0	3.5	7	28	35

② **a** $x = 0.87y$
 b £22.62
 c 22.70 euros or £19.75

③ $s = 0.3r$, $r = 500$

④

U	0.5	1	1.5	2	3
V	12	6	4	3	2

⑤ **a** $\dfrac{1}{2}$ hour   **b** $1\dfrac{1}{4}$ hours